Adventures in Science

From Quantum Thinking to Alien Encounters

Adventures in Science

From Quantum Thinking to Alien Encounters

Mt. San Antonio College
Walnut, California

First Printing: 2014

ISBN: 978-1-56543-804-0

MSAC Philosophy Group
Mt. San Antonio College
1100 Walnut, California 91789 USA

Website: http://www.neuralsurfer.com

Imprint: *The Runnebohm Library Series*

Dedication

To our two boys:

Shaun-Michael and Kelly-Joseph

Table of Contents

Acknowledgements

Andrea and I would like to express our deepest thanks to Frank Visser and *Integral World* for publishing a large number of our articles over the years and for encouraging us in exploring the frontiers of neuroscience, quantum physics, and evolutionary biology.

The MSAC Philosophy Group

MSAC Philosophy Group was founded at Mt. San Antonio College in Walnut, California in 1990. It was designed to present a variety of materials--from original books to essays to websites to forums to blogs to social networks to films--on science, religion, and philosophy. In 2008 with the advent of print on demand and cloud computing, the MSAC Philosophy Group decided to embark on an ambitious program of publishing a large series of books and magazines. Today there are well over 100 distinct magazine titles and 50 book titles. In addition, the entire MSAC database is now being put online via Amazon's Kindle, Barnes and Noble's Nook, Google's eBooks, and Apple's iBooks. A special mobile app called Neural Surfer Films is now available for Apple's iPhones and iPads, as well as one for Android operating systems on smart phones and tablets. *The Runnebohm Library* contains works on Einstein, Turing, Russell, Crick and other luminous thinkers. Some of the more popular titles include, *Darwin's DNA: A Brief Introduction to Evolutionary Philosophy* and *Global Positioning Intelligence: The Future of Digital Information*. Finally, *The Runnebohm Library* is in the process of producing a number of highly interactive texts that will include embedded video, games, and interactive feedback loops.

1 | *Shiva Science*

"Science is essentially an anarchic enterprise: theoretical anarchism is more humanitarian and more likely to encourage progress than its law-and-order alternatives."
--*Paul Feyerabend*

"The scientist is not a person who gives the right answers, he's one who asks the right questions."
-- *Claude Lévi-Strauss*

We are trying to prove ourselves wrong as quickly as possible, because only in that way can we find progress."
-- *Richard P. Feynman*

It is one of the curious oddities of our time that we talk so much about the scientific method as if it is one singular entity when, in point of practice, it is anything but.

The ultimate linchpin in science is decided not by how we go about doing it, but about how well our hunches, observations, and results tally with the universe we observe and, in turn, how such intellectual lurches compare and contrast with other competing stratagems in terms of yielding more, not less, information.

The very word science, derived from the Latin "scire" (see the *Oxford English Dictionary* for more on its etymology), is rather open ended and simply means "knowledge." Hence, it is along these lines that the acerbic philosopher, Paul Feyerabend, long argued that science isn't really a method at all in the strict sense of that term. Rather, it is a label we use to describe the testing and verifying of differing ideas and maps we have about the world and allowing such templates to strenuously compete with one another. How we actually do science is invariably much sloppier than we might at first suspect, but which every scientist worth his or her salt knows too well.

This is why Richard Feynman spoke so frankly when he said that the first thing we do when we try to figure out how something works in terms of physical laws is by "guessing." We then check to see how well our particular guesses hold up.

Given this modus operandi, we soon realize--along with T.H. Huxley--that we are all scientists at some level, whether it is trying to figure out how to bake a cherry pie for Christmas day or fixing an old pool heater that has seen better days.

Yet, sometimes certain authors in their haste to criticize some reductionistic aspects of science resort to misleading caricatures about such an endeavor, neglecting in the process how truly accessible (and messy) the scientific endeavor can be.

For instance, in a recent article ["In Defence of Ken Wilber's Integral Theory of Everything"] published on Integral World, H.B. Augustine opined that "Science is destructive in character because it serves to disprove and criticize," which, of course, is neither accurate or insightful.

Astronomy, to take just one obvious example from the sciences, isn't so about disproving anything as such, but rather about discovering vast new vistas. There is nothing destructive per se when Edwin Hubble looked through his telescope in the latter part of the 1920s and observed a red shift in the electromagnetic spectrum, whereby light from distant stellar objects appeared to moving away from his point of inquiry. Hubble's record only served to confirm what the Belgian priest and physicist, Georges Lemaître, had deduced two years prior from Einstein's general theory of relativity which is that the universe must be expanding.

Yes, it is certainly the case that reductionism is part of science, but so is almost any human endeavor, including the very use of symbols (from Mandarin Chinese to Sanskrit) and vocal sound waves to communicate and explain complex features found in nature. But to suggest as Augustine naively does that "the scientific method is all about 'divide and conquer'" is to overlook the fantastic panorama of how we go about gathering knowledge.

Contrary to popular misconceptions, there is nothing holding any would-be scientist back from taking a wholistic or integral approach to his subject. Indeed, many great theorists

2

have already done exactly that, ranging from such Nobel laureates as Ilya Prigogine who focused on complexity and dissipative structures to Gerald Edelman whose pioneering work on the brain and consciousness is underlined with a deep appreciation for the emergent properties of qualia and self-reflective consciousness which he refers to as second nature.

> Sometimes things can be explained by studying their constituents—sometimes not."
> — *Steven Weinberg*

The noted physicist Steven Weinberg points out that a distinction should be made between grand reductionism and petty reductionism, since the latter, he argues, is "not worth a fierce defense. Sometimes things can be explained by studying their constituents - sometimes not. When Einstein explained Newton's theories of motion and gravitation, he was not committing petty reductionism. His explanation was not based on some theory about the constituents of anything, but rather on a new physical principle, the general principle of relativity, which is embodied in his theory of curved spacetime. In fact, petty reductionism in physics has probably run its course. Just as it doesn't make sense to talk about the hardness or temperature or intelligence of individual "elementary" particles, it is also not possible to give a precise meaning to statements about particles being composed of other particles. We do speak loosely of a proton as being composed of three quarks, but if you look very closely at a quark you will find it surrounded with a cloud of quarks and antiquarks and other particles, occasionally bound into protons; so at least for a brief moment we could say that the quark is made of protons."

Indeed, the renowned Santa Fe Institute, founded by such luminaries as Murray Gell-Mann (Nobel prize winner for his theoretical work on elementary particles and more popularly known for his naming of "quarks"), supports a whole host of "non" reductionistic scientific studies.

Thus, it is ironic that in prematurely trying to criticize a particular aspect of science (among a host of varying other aspects) one would resort to "reducing" all of science into a cheap and wholly misleading sound bite in order to champion its purported alternative.

No, the truth is that the human drive to gather knowledge and then openly compare and contrast such theoretic and observational templates is an expansive field and one which allows for all sorts of imaginative pathways. In other words, science is a quest with many methods and not just one.

Analogously, science is like the Hindu god Shiva, who is often depicted with multiple arms signifying his illustrious powers to take on different forms and act in manifold ways. While it is certainly true that one aspect of his nature is destructive, it is also true that Shiva is depicted as a transformer. Accordingly, Shiva has numerous aspects ranging from beneficent to frightening. But one of his more popular aspects is as a universal dancer, perhaps best illustrated artistically in bronze casts as Nataraja.

Science like Shiva cannot be confined to only one aspect. To focus on only one feature of science to the exclusion of others in order to misleadingly characterize its multiform nature (so as to elevate another method to a higher ranking) is to perpetuate a falsehood.

Ironically, even though we value science for many of the predictions and technological inventions it makes from time to time, we imprison its potential value when we try to limit its Shiva-like nature. As Lewis Thomas insightfully pointed out in his 1980 tome, *Late Night Thoughts on Listening to Mahler's Ninth Symphony*, "Science is useful, indispensable sometimes, but whenever it moves forward it does so by producing a surprise; you cannot specify the surprise you'd like."

There are, in sum, innumerable ways to gather knowledge about the cosmos we inhabit. What makes science so powerful is that it allows for such findings to be publicly aired and scrutinized and tested, not only with the world it is trying to understand but with other vying alternative models which differ from each other. Perhaps science's greatest contribution is that at its best it is open to refutation and is thereby open to change.

Carl Sagan summarized it thusly, "In science it often happens that scientists say, 'You know that's a really good argument; my position is mistaken,' and then they would actually change their minds and you never hear that old view from them again. They really do it. It doesn't happen as often as it should, because scientists are human and change is sometimes painful. But it happens every day."

An often-told joke concerning Albert Einstein perhaps reveals best the tentative nature of science and its findings: *Student:* "Dr. Einstein, Aren't these the same questions as last year's physics final exam?" *Dr. Einstein:* "Yes; But this year the answers are different."

2 | On Reductionism

It seems to me that we often get caught up in a tripartite dilemma—one that is echoed in the following intellectual triangle:

pretext, text, and context;
or prehension, apprehension, comprehension;
or deflation, elation, inflation;
or reductionism, phenomenology, complexity.

Take a book, actually any book, but in this example we will simply limit it to that wonderful little text published by Cambridge University Press, *What is Life?* by Erwin Schrodinger. Not very long (my edition is only 96 pages), but sufficient to get the point across.

Now if I wanted to know the meaning of the book, if I wanted to "apprehend" its contents, I would read the entire tome. So far, so good. But let's say that I wanted to discover what ultimately constituted the book. That is, I wanted to know what the common underlying symbol system was that actually comprised the text. In this type of quest (the pun is intentional), I would reduce the book down to its chapter divisions, chapter divisions down to pages, pages down to paragraphs, paragraphs down to sentences, sentences down to words, and, finally, words down to letters.

Letters, individual (but only 26 variations in our alphabet) symbols would be the fundamental unit by which information is encoded. But let us imagine that I only read *What is Life?* in terms of its letters, not the words they form, or the sentences they create, or the paragraphs they construct, or the pages they comprise ... What then? Would I "understand" the meaning of the book if I simply limited my reading to the symbol units themselves? The answer is fairly obvious: no.

Why? Because the letters "a" or "b" or "z" do not, indeed cannot, convey the meaning as isolated units. They begin to form meaning when they conjoin and develop a larger complex, a larger construction.

Okay, so now we understand something called "pretext." That unit which is rudimentary to the book, but which is not yet readable as a text. Important sidebar: as the latest studies in reading have shown, "hooked on phonics" (Michael Landon's greatest legacy?) or the understanding of the sounds that constitute words is very helpful to students for later reading and comprehension. Indeed, as good as "wholistic" reading may be, phonics is even more helpful at a fundamental level. The reason why is pretty obvious: the more one grounds herself in "pre-text" the more secure the formation will be when one moves "up" to "textual understanding." Why? Because there will be less *confusion* of word or sentence formation. Pretext is fundamental.

But now let us imagine that we have understood pretext (symbols, or the alphabet in our example) and text (the larger complex which words and sentences and paragraphs develop). Is that all that is necessary to "comprehend" (I am consciously using a different word than "apprehend" at this stage) Schrodinger's book, *What is Life?*

Well, yes and no. Yes, because clearly I can get a fairly decent sense of what our quantum theorist is trying to convey. No, because there are certain things "around" or "beyond" the text that the book cannot convey, but which strangely enough is demanded for better comprehension. That missing something is not pretext, or text, but context.

Context can range from the very simple: the paper quality, the binding, the typography, the smell of the book (anyone who has read books published in India will immediately know what I am talking about) to the very complex: What year was the book published? What prior knowledge of math, of physics, of astronomy, of biology is necessary to better engage the text? Moreover, what is my mood? What country am I in? What religious/scientific beliefs do I bring to the text? Is it nighttime? Is the T.V. on? Silly questions? Not really, since this larger infusing environment—ranging from the rudimentary to the baneful to the sophisticated—plays an important factor in any reading of any book at any time. Context is the larger arena by which any text, formed by any pretext, is understood.

Pretext: rudimentary/fundamental;
Text: instrumental/informative
Context: bounding/eliciting/forming

And, as such, a pretext can "evolve" itself into a text which in turn "evolves" itself into a context ... So that a context in a different situation can become the pretext to a new text which itself is housed by a new context ... And so on and so on. Or, as Ken Wilber would have it, "holons, holons, and more holons." Each holon, as Wilber would suggest, comprised of holonic parts, but which itself acts as a holonic part (I know it sounds oxymoronic but that's the point) of some larger holon.

Atoms have parts (electrons and protons, for instance), but an atom is "part" of something larger (molecules). Molecules have parts (atoms of varying weights).j, but a molecule is "part" of something larger (cells). Cells have parts (molecular bonding), but a cell is "part" of something larger (a simple organism). Etc. Which leads to brains have parts (neurons, axions, synapses....), but a brain is "part" of something larger (the human body). The human body has parts (the brain, the heart, the liver....), but the human body is "part" of something larger (family/environment) ... and so on.

Okay, so who cares? Well, it seems as if we are always struggling with this dilemma: reduce or inflate? Pretext or context?

What Wilber suggests, though I don't think he is aware of how thoroughly materialistic his system can be, is that science tends towards reductionism, towards Occam's razor, towards Churchland's intertheoretic explanations, and that is its great strength ... It tends to explain things more simply ... and by more simply we mean more "fundamentally."

For this reason it usually kicks butt on any or all "inflated" theories (context which seems divorced from pretext?). But reductionism can in fact "go too far." And what I mean by "too far" is that it can actually become anti-informational when such reductionism to echo the words of Dennett becomes "cheap."

What is cheap reductionism? Let's go back to our book, *What is Life?* Remember that we can reduce the text to a series of letters, and such reductionism would be very helpful at first. Indeed, it would give us a tremendous grasp of what we could or perhaps could not do with such a symbol system.

But let's imagine that I wanted to "reduce" even further. I find that the letters are made of little lines, so that a "W" looks like two "V's (VV) conjoined. We can even go farther and see that all letters in print are made of tiny molecules which are themselves made up of atoms, and the atoms are (eventually) made of quarks ... and quarks? Well, super-strings in nth-dimension are vibrating at a frequency below Planck's constant, which is not at any perceptive level.....

See, we have gone too far. It is nice to say that words are nothing more than the congealed results of trapped electrons but it adds very little in terms of instrumental or pivotal information. To be sure, it helps to understand other things about our universe, but given the complexity of letters it does not add to our present domain-related discussion.

We have made a classic boundary skip, category collapse, and we have indulged in "cheap" (not very worthwhile) reductionism. Yet, this does not mean that reductionism is bad, it just means that reductionism is quite useful in the right domain ... Reduce too much and we lose. Don't reduce and we inflate too much.

Having said all this, it is obvious that the direction of science must always be, fundamentally, in the path of "reducing" as far as it can go while remaining useful and informational.

Let us see how this works across disciplines: physics (looks to math, perhaps our most precise and accurate "human" language); chemistry (looks to physics, especially quantum mechanics—just see the work of Linus Pauling); biology, especially molecular biology (looks to chemistry; just think of Watson and Crick and the double helix model of DNA); psychology (looks, or perhaps they should!) to neuroscience, the study of the brain (Why? Because the greatest progress in psychology has not come from Freud or Jung.... It has come from those pioneers who have grounded their studies in evolutionary brain science....).

Just think to yourself when you get a headache: should I call 1-900-shaman or should I take an Excedrin extra strength; sociology (looks bad, but if it is to have a future, it should tend towards biology, psychology, evolution). I know nobody likes Sociobiology (its "new" name is evolutionary psychology—cool, now we can do what E.O. Wilson has been talking about for some twenty years). A sociology grounded in evolutionary theory may actually come up with some revolutionary and cross-cultural predictions. Get rid of Marx, but not his reductionist spirit, and hitch it to evolution and sociology may go somewhere.

Now each of these disciplines "succeeds" when they discover their basic alphabet, their basic pretext. That has to be done and is the key to any further scientific progress. Below is a very simplistic look at various pretext/alphabets: physics: at the subatomic level it is clearly quantum mechanics, and its sophisticated reworking Q.E.D., which is so precise that no theory rivals it in terms of accuracy. Molecular biology: DNA as everyone knows is the blueprint for all life. Evolutionary biology: natural selection. Darwin's contribution, as Dennett so rightly points out, is probably the single greatest thought given to man.

But let's go back to *What is Life?* Can I get a Q.E.D. reading of it? Can I get a DNA reading of it? Can I get a natural selection reading of it? Yea, but it is not going to be as useful or as informative as understanding the English language at its own level. So let's jump domains and cut to the chase: Metaphors coming (literalists beware!) 1. The brain is a text 2. Neurons are its subtext 3. The human body (thanks to *Descartes' Error* for informing us here) is the context.

Now I want to understand "me". Best to start with the brain's alphabet (neuroscience 101); Then look at the brain's architecture and how the neural symphony comes together And then look at the larger human anatomy and see how all the various "beyond" brain parts work together.

Is that enough? No, because just as neurons have constituent parts, so does the human body have a much larger context or field of interplay. That larger context or infusing environment always arises when the pretext and text have reached their terminus, their limit. Thus if we want to avoid "cheap" reductionism, we also want to avoid "expensive" inflationism (think of "fake" or "monopoly" money).

Why did I go on at length about pretext, text, context (alphabets/ books/ purviews)? Because in trying to understand consciousness (soul?) we usually run in two directions: deflation—hey, Crick's right, I am nothing more than a sophisticated neural net; or, inflation—hey, Wilber's right, I am Pure Spirit.

In the middle of such opposing sides is relation: the connection between these two views (such as, hey, all I know is that I "feel" more than the body, except in those cases when I have a really bad headache or toothache.).

What is fairly obvious in understanding a book (pretext: alphabet, phonics): *text*: words, sentences, paragraphs, chapters; *context*: When was this book written? Where was it published? What mood am I in when I read it?) can also be applied via analogy (literalists beware!) to consciousness: pretext: brain, neural net, connectionist, PDP; text: "I" consciousness, personality, "The lived through sense of me" context: in what city does this "I" live; what relationship do I have to my family, to my nation, to my religion, etc.

Given this simple scenario it becomes obvious that we can reduce consciousness down to its pretext (the brain) and we would be only partially correct. We would not—perhaps could not—understand the "qualia"—the phenomenology of my own lived through experiences (John Searle's "first person") if we merely stayed at the level of neurons. No doubt, we would understand a tremendous amount (and my biases lean, I should point out, with the Churchlands' for intertheoretic reductions whenever possible), but something would be lost in the reductive translation. We need text (read: the personality at its own level, at its own understanding, at its own self-reflections)

Moreover, there is something about consciousness that is not merely the brain, but also the body entity (as *Descartes' Error* strongly suggests). Additionally, consciousness, as such, arises within a larger field, that of family relations, societal relations, ecological niches, etc. It is this larger field which informs and shapes much of what we know about our consciousness and personality. This larger "context" is essential, especially when one considers the vast differences in cultures throughout time and place on this planet.

The above tripartite schema is clear enough and I would venture to guess that most would not disagree with it. Where we run into difficulty is when we start to think of consciousness as "transcending" physicality. Well, to be sure, there is a transcendence of sorts when the alphabet turns into words and words into sentences and so on. But it is not divorced from the prior structure. Indeed, each higher level is situated upon—sits upon—that former and under girding pretext.

Okay, *The Great Gatsby* transcends a mere random collection of letters (there is a point, there is meaning, there is character development), but take out those very letters at any stage and the entire superstructure of the "novel" collapses.

As Wilber would point out (or any good physicist for that matter), the alphabet is more "fundamental" than sentences, though sentences are more significant (convey more meaning, have more depth). So at each stage of explanation we are confronted with this situation: what is the pretext? (alphabet, the rudimentary symbols by which we comprise larger sets. Hint: this can be applied to anything: Atoms? Electrons/nucleus. Molecules? Atoms. Living Cells? DNA ... and so on)

What is the text? (This is actually quite arbitrary and it depends where and when we want to measure something, but once staked out it becomes the rallying point for pretext and context) what is the context? (In what larger field does the alphabet, the DNA, the atoms, the quarks, etc., arise?)

But here's the catch: none of these larger texts or contexts is divorced or separated from its predecessors. Indeed, in terms of genealogy, it is impossible to have a book, as such, without a rudimentary symbol system. It is impossible to have molecules without atoms. It is impossible to have a brain without neurons (A note of caution to my A.I. friends: this is merely an analogy; I am not denying that silicon chips could not in theory replace neuronal components .... Even then, there is still a pretext— sand!).

13

So when one speaks of consciousness without a brain, or beyond the body, or without physicality, it is naturally criticized by those conversant with neurology. They don't buy it, since they know that by understanding the pretext of the brain they can actually change how the brain functions. They know the code. And there is nothing to suggest that consciousness which arises in the brain can somehow fly away from the body or code without any restraint whatsoever.

But this is exactly the point about any physical or mental or spiritual thing—things arise from other things and those very things arise in fields of emergence. Yet, there is no absolute separation from the quantity of one thing into the quantity of another (or new thing).

To put it to consciousness, we have the following: neurons: subset brain: set personality: post-set or personality: preset society: set ecology/environment: post-set all the way down, as Wilber says, holons all the way up. Yet, Wilber makes one huge "sky-hook" mistake (thanks Daniel Dennett) when he argues that Spirit is the basis of all matter. Wilber wants us to believe that Spirit is not based upon matter, but the reverse.

This is where he makes his leap and where any materialist worth his salt is going to have huge difficulties with Wilber. What Wilber should concede (he doesn't) is that he does not know what Spirit is ... (I don't either). Why? Because what Wilber really means by Spirit is the Context of every pretext/text/context. That is the Infinity in which everything arises.

Well, I don't know what that is; Wilber does not know what that is; I would imagine nobody "knows" what that is. What we do know, partially, are limited frames of reference, and, as such, we can pontificate upon them—from quarks to atoms to molecules to cells to people to societies to nations.

But let's not go too far. There may be an astral plane, but we have no evidence—at this stage—to comprehend it. We only have limited symbols that may point to it. Yet, do we admit to this contextual impasse? Do we, in fact, say with humility, "there might be?" Yes and No.

When one reads Wilber or anybody (including almost all of my early writings) you get the impression that he/she/it has a lock on the ultimate truths. Woe, we just found out about DNA ... and that only explains the alphabet of life processes. Before the 50s we didn't know. So can we then take a huge leap from DNA to the fabulous inner regions of *Sahans-dal-Kanwal?*

We can, but my hunch is that we are merely "infusing" contexts that we do not as of yet know exist. And by doing such we "confuse" ourselves unnecessarily.

Dennett would say we are going for skyhooks, not cranes. But it is cranes all the way up that produce the higher orders, the complex systems ... Not the other way around. What meditation may indicate is a higher context but that very higher context will, by necessity, be grounded in the text that precedes it. What is that? The brain. So there is no way around this— from a bottom up perspective—but to admit that everything is higher order materialism. I could say "spirit" but that would incline itself to meaningless gibberish.

When I say "matter is all there is," it tends towards reductionism and thus is more locatable in terms of its pretexts. It does not mean, of course, that I "know" what matter ultimately is. I don't. What it suggests is that we ground our speculations, as always, with the rudimentary tools that are available. Wilber and others with a Consciousness bent (read: *Big Context Takes Over All*) will heartily disagree with my slant, thinking that I have sold my soul in exchange for quantum mechanics and neurology and evolution.

No, I have come to grips with the fact that whatever my soul may be is grounded in the pretexts/texts/contexts of everything that arises within my body and without it. Thus I think we will understand a lot more of what we mean by "soul" if we start with what we mean by "body," by "brain," etc.

Now if there is indeed a context that transcends this frame of reference, this waking state, then having a humble approach to it, I would argue, will be even better. We will be more open and skeptical. Open to the possibility; skeptical of misguided confusions of pretext and context. What we get instead with Wilberian type thinking is actually a bit of hubris and a whole lot of arrogance (well, he is on the seventh plane.... she is on the eighth plane.... and Free Willy is stuck in the bowels of the astral).

Think for a second. How many times have we engaged in arguments over inner level attainments from the comfort of our own chairs? And all the while most of cannot even explain—in terms of physiology—how we grow hair on our arms.

I find it completely amazing that we cannot define this universe (which is the emerging context by which we can empirically understand our lives), but geez we most definitely know the ins and out of the astral plane and who resides there.

Materialism is multi-dimensional. The reason it is probably nicer to use that term (versus spirit) is that it tends towards reductionism. It tends towards looking to its primary alphabet.

By positing a materialist position, we look first for the cranes; by positing a spiritualist position, we look first for the sky-hooks. Or in the terminology I have been using: by saying matter first, we look for pretext; by saying spirit first, we look for super-context (usually divorced from any empirical referent)

Now it is correct to look for both pretext and context, but both arise in a relationship in the here and now. Super-contexts (or Spirit) is usually too far up the hierarchy to mean anything useful. I could say anything to posit a super-context (Elvis, Sugmad, Anami Purush, God, Jesus, Gumby, etc.), but it would given our grounded lives here mean zip.

Why zip?--because it can stand for anything. Yet matter is the same thing, one might counter. Yes, but it has one advantage: it looks to its predecessors, to its genealogical parents, to its alphabet, to its cranes, to its roots ... And when such is lacking, then and only then can it make the slow climb up.

Spirit—we could actually say Infinity instead—is never definable because it is by definition never ascertainable or limited. Matter is about limits or at the least our understanding of how energy/matter interacts at a certain level (our text?).

But matter is not something flat, not something grey. It is absolutely beyond my fullest comprehension. And yet, as Einstein rightly said (he was agnostic and a materialist) "the most incomprehensible thing about the universe is that it is comprehensible." Thus we may not ultimately know, but we know a little and that little grows by looking to its roots.

3 | Mysticism's Version of Intelligent Design

India, Punjab, Dera Baba Jaimal Singh, December 1981, Evening, Guest House. Questions and answers with Maharaj Charan Singh, Spiritual Leader of Radhasoami Satsang Beas.

It was a serene evening and all the foreign visitors at the Dera from various countries around the world, but particularly the United States, the United Kingdom, and Brazil, were gathered at the Dera Guest house's upper meeting room to have an evening session of questions and answers with Charan Singh, the spiritual leader of Radhasoami Satsang Beas.

I remember that the discussion turned to science and evolution and at one point in the proceedings it seemed as if Charan Singh was positing a creationist position. I raised my hand and explained to my guru that I taught evolution in my Chaminade College Preparatory's classes and went on to say something to the effect about whether I should be more nebulous about it, given his own questioning comments on the subject. Charan laughed and then proceeded to say that science should be taught in science classes and that in any case it was a mystery beyond our mental comprehension. I think Charan also commented that maybe the monkeys wouldn't be happy knowing about our relationship with them (reversing the usual critique of how humans don't like being related to primates).

I wasn't rankled at the time by his answer, since he didn't categorically dismiss Darwin or evolution by natural selection, even if he did suggest a more mystical interpretation of the how varying species came into being.

However, I did become perturbed eight years later at the Dera where I got into a relatively heated discussion with a fellow satsangi and author John Davidson when we were standing in line to do our laundry. Our conversation started off friendly enough as he was working on a book about Christianity and was keenly interested in early Gnosticism. However, and I am not quite sure how the topic came up, we soon started talking about evolution and I was wonder struck by Davidson's own creationist leanings. I said something to the effect that his ideas were not evidential or scientific. I also asked him if he had ever received criticism for what he had written and I was a bit flabbergasted by his apparent unwillingness to have his theories critiqued by others more versed in the subject. Davidson then tried to shore up his argument by saying that Charan Singh and apparently all the other masters in the Beas lineage, held a view similar to his own. I then said something a bit impolite (given the surroundings) about how our guru really didn't know science very well and therefore shouldn't be cited as an authority on the subject. Well, our conversation went from bad to worse and finally Davidson tapped me on the nose to silence my increasingly vocalized skepticism of his nuanced version of intelligent design. I don't think we talked much after that.

Over the years I dismissed Davidson's books on how the universe and life came into being and related subjects as mostly New Age silliness. It didn't occur to me that many of his peculiar views on creation would later be rehashed and published by the Radhasoami Beas organization. I must confess that I was a bit surprised and disheartened to see Davidson's deeply questionable perspective given the apparent imprimatur of the "Science of the Soul Research Centre", particularly since his views are anything but scientific.

The following critique therefore is aimed both at Davidson and Radhasoami Beas since both seem to behold similar views on creationism and evolution. Given that Radhasoami under the direction of its current leader, Gurinder Singh, wants to advertise itself worldwide as a scientific enterprise, even if of the spiritual variety, it must be more open to critical views from both inside and outside of their sangat. Science is after all predicated and built upon that most necessary and highly treasured feature that permeates its entire edifice--the ability to question and to doubt.

THE REVIEW

Quite frankly, *One Being One* is a dishonest book written not so much to bridge the gap between science and religion, but to not so subtly proselytize for Radhasoami's version of intelligent design.

The first problem we confront with Radhasoami's attempt to advertise itself under the all encompassing banner of "Science of the Soul Research Centre" is figuring out what exactly they mean by science. In the Publisher's Note to John Davidson's book it states, "*One Being One* presents a reconciliatory perspective within the current dialogue between science and religion. It is hoped that it will contribute to a better understanding of these two fields, which are often presented as contradictory."

In his preface to his tome, Davidson explains that he has had a life-long interest in "nature, science, and mysticism." He mentions that a week after getting initiated by Charan Singh into the Radhasoami faith in 1967, he got his "first real job at the Department of Applied Mathematics and Theoretical Physics at Cambridge," where he apparently assisted others in computing which he states "became my profession."

Davidson is very clearly attempting to show the reader that he has a background both in science and mysticism and because of this is well suited to guide us on the "relationship between scientific and mystical traditions." As I will try to show, however, Davidson's understanding of science is fraught with difficulties. At one end, he wants to argue that Radhsoami's version of mysticism is a scientific endeavor on par with physics and chemistry, and on the other end he wishes to severely criticize science's naturalistic methodologies. In doing this, it becomes readily apparent that Davidson's overall agenda is to have science fit in with Radhasoami's mystical worldview and when it doesn't he invariably finds fault with science but he never (not once) questions Radhasoami's obvious theological purview.

Thus he is not reconciling science with mysticism, but rather championing Radhasoami's viewpoint above all else. Although he might be loathe to admit such a comparison, Davidson's apologetics are not dissimilar to how certain Christian fundamentalists cherry pick science to support their Biblical worldview. Yet when the science doesn't dovetail with their belief system (as in the case of Darwinian evolution), they dismiss it and then pontificate on the superiority of what the Bible says about creation. Ironically, this is almost precisely what Davidson does throughout his book, except that he elevates mysticism (and not the Bible) at each and every turn.

Davidson laments on page 11 of his book that "modern science suffers from the lack of an inherent spiritual perspective" and goes on to categorically allege that this absence has led science down a "road of self-destruction" which have "led to the rape and wreck of our planet in a previously unprecedented manner," a claim he will later contradict on page 172 when he writes, "It's not science itself that is to blame for all the environmental havoc."

Contrary to Davidson's lamentation, science is a process of discovery and succeeds best when it is not weighed down by metaphysical considerations. Indeed, it is precisely when science got divorced from religion that it made its greatest progress. Moreover, what Davidson takes to be spiritual is not necessarily what a Christian or a Muslim or a Buddhist might mean by the term.

Davidson also mistakenly limits the import of what science can potentially achieve when he misleadingly claims that "science as a logically consistent description of the mechanics of the material universe rigorously excludes [my emphasis] anything but raw empirical data and intellectual theorizing thereon."

This is simply untrue. Science doesn't exclude anything a priori and there is nothing whatsoever holding any scientist (professional or amateur) back from exploring anything that strikes his or her interests. For example, there have been pioneering studies done on consciousness at M.I.T., Harvard, U.C. San Diego, and other top notch hard science universities. In addition, for decades studies have been conducted at Duke and Stanford on such paranormal subjects as E.S.P and telepathy. More recently, meta studies have been conducted on remote viewing and astrology. Several universities have even funded research into out-of-body and near-death experiences. Thus, Davidson's sweeping generalization overlooks what scientists are actually doing.

In science one can proffer any sort of guess one wishes about how and why a certain phenomena occurs. The key, however, is that science demands that such guesses eventually be tested to see how well they explain a given object or event. If they fail to live up to their hyped billing, then the scientist must be willing to admit to their insufficiency and try to find a better hypothesis which explains the data more accurately or more comprehensively. Interestingly, it is because science and scientists must be willing to be proven wrong which has led to the explosion of our current knowledge about how the world works.

Yet, the real thorn in Davidson's side appears to be the theory of evolution by natural selection as first developed by Charles Darwin and his later intellectual heirs such as Julian Huxley, R.A. Fisher, Theodosius Dobzhansky, Ernst Mayr, George Simpson, W.D. Hamilton, E.O. Wilson, James Watson, Francis Crick, and others.

It is here that Davidson echoes his earlier books on creationism and evolution, none of which have received any serious consideration by mainstream science and for good reason. He wants to reintroduce, even if hesitatingly. a variant version of Intelligent Design but not have it hitched to the Bible. As Davidson writes, "That makes no sense to me [connecting Intelligent Design to Genesis], but the idea of nature being permeated by an essential Intelligence or Consciousness that gives rise to all the natural 'laws' and 'principles' known (and unknown) does not seem like such a bad idea."

The powerful thing about science is that one's personal preferences (whether one finds Biblical Intelligent Design or Mystical Intelligent Design persuasive) are besides the point since what is ultimately important is whether a theory is evidential or not and how well it can withstand critical scrutiny. In other words, what may make no sense to one person may make lots of sense to another. But in either case such hunches mean nothing until they are tested and that is where science comes in. Yes, we may have all sorts of hunches and predispositions about how nature "should" work, but as Richard Feynman, the noted physicist reminds us, science is only concerned with discovering how nature does indeed operate, sans our own biases. And in this way, we must have a never-ending ability and openness to being wrong. This is primarily why Karl Popper's notion of falsification has been so widely cited as a way of differentiating potentially viable scientific ideas from merely pseudo-scientific claptrap. We must not fool ourselves and we are the ones that can be most easily duped. Science is a form of systematic doubt and successful experiments in science are those that can withstand doubt (as in rigorous testing, etc.) from varying quarters best.

Yet, does Davidson or Radhasoami Beas, for that matter, posit the same skepticism of their own mystical traditions as they do with evolution or any scientific claim that contravenes their "a priori reasoning"? In Davidson's book we never see him raise even the tiniest of doubts about mysticism and its methods. Why is this so? Why be skeptical in one arena but not the other?

Science cannot diminish the spiritual quest if that quest is concerned with truth and not merely dogmatic assertions taken at face value. However, spiritual paths must be open to varying interpretations and must (and this is the kicker that most paths tend to resist) be open to falsification. In other words, for any endeavor to be justifiably regarded as "scientific" it must be willing to be corrected, to be changed, and to be wrong.

While much of spiritual literature advertises itself as scientific, we find that the very basis of almost all scientific endeavors, that of theory making, is dismissed since it interferes with certain strongly held religious beliefs.

For example, when a particular guru instructs the neophyte to go within the laboratory of one's own body to verify the factualness of their respective theology, he doesn't mention that such subjective experiences should be open to varying interpretations of what they could possibly mean or how they may be a byproduct of the brain.

This version of science is more akin to an elaborate food recipe, where the would-be chef needs to follow a set of given instructions in order to know how to make a chocolate cake or a vegetarian pizza. Applied science is successful only after certain facts are well established and known. But at this juncture, any proposed spiritual science is still in its infancy, even if some traditions would like to suggest otherwise.

Such versions of science are similar to a computer program like Basic or Unix where if you follow just the right set of protocols you will invariably end up with a repeatable outcome. But this leaves out the most vitally important aspect of science, something which only few meditational teachers seemed to grasp, which is that science isn't a thing but rather a process of discovery and along that pathway there will be false starts, differences of opinion, falsifications, tentative hypotheses, and theories and even facts that are always subject to alteration or even wholesale elimination.

For instance, it may be one thing for a religious tradition to say, "There is no better method than that of the Sound Current, which is an ancient and natural science. It was designed by the Creator Himself, is within every one of us, yet whole nations and entire countries of the world are ignorant of it." But, it is quite another thing to then claim that such a description is part and parcel of a genuine science. Notice that the preceding quote concerning shabd yoga isn't alleging to be merely a scientific endeavor to be placed alongside biology or chemistry, but rather is emphatically stating (quite categorically one might add) to be more fundamental than even physics since it was created by God himself as a path back to him. While a devotee may believe this to be the case, it is fairly obvious to an outsider that this assertion is not a scientific claim as much as a dogmatic one in the guise of scientific dressing.

This is important to understand since a genuine scientific endeavor worthy of its name cannot arbitrarily pick which aspect of science they wish to utilize. In cases like this it is as if science is employed as a form of advertising to reach interested seekers who may have been turned off by more exclusive forms of religious dogmatism. While it appeals to the rational authority of science, it does so by claiming that shabd yoga is the highest of all sciences, apparently forgetting in the process that any scientific endeavor worth the appellation must be open to revaluation and correction. Nowhere do we find in most meditational literature a consistent theme of falsification, where past gurus and their ideas are corrected, changed, or overthrown. What we do find, however, is a paradoxical selection of quasi scientific language which appears to be offering a potential experimental procedure to validate inner spiritual experiences.

But as we have previously noted, when closely examined this type of rhetoric is more an instructional formula to achieve an already agreed upon result (similar to baking a pie) and less a scientific method with all its unforeseen trajectories.

Surprisingly, shabd yoga or any meditational discipline that wishes to be viewed as a genuine science would be best served if, instead of first resorting to dogmatic axioms about its ultimate truth claims or appealing to unassailable authorities in its lineal past, it looked for ways to falsify itself. A good example of how to do this can be found in Charles Darwin's *On the Origin of Species*, where in just one sentence he explained how his whole theory of evolution by natural selection could be wrong. Wrote Darwin,

"If it could be demonstrated that any complex organ existed, which could not possibly have been formed by numerous, successive, slight modifications, my theory would absolutely break down."

Here Darwin points the pathway by which his whole theory could be turned upside down. Any spiritual endeavor desiring to be taken seriously as a science should do the same.

One of the chief methodological problems with Davidson's book (as he readily admits on page 14) is that it "relies on a priori reasoning, on taking certain things for granted. I've tried to present things in a logical manner, but I do presume the existence of an all-persuasive, unseen Being or Intelligence."

With such a metaphysical bias running throughout his book, the reader knows right from the outset that he or she is not in Kansas anymore. Regretfully, Davidson doesn't stop there but tries to equate his speculative reasoning with scientific naturalism (which grounds itself in the empirical universe around us) incorrectly suggesting that they are the same. Lest Davidson ignore the obvious, science first focuses on correlations and causations that can be verified in this world. But it doesn't claim, as he wrongly infers, that there cannot be something that transcends terra firma. Science, unlike Davidson's mysticism, doesn't have such an over-arching transcendental philosophy. Here Davidson has mistaken science's practicality with an underlying metaphysic.

Take for example the issue of whether science can indeed decide the physicality of consciousness, even despite its powerful tool of methodological naturalism. Quite simply, if consciousness is indeed beyond physics or anything within its known laws, then no matter how hard we try to ground mind to its neural structures there will always be something missing in such reductions. And, interestingly, this gap will loom even larger because our physical science will be unable to adequately explain it. Thus, one could argue that such a physicalist approach will shine a much more illuminating light upon the problem by showing exactly where, when, and how awareness is not the result of physical properties. But if we forego this grounded scientific quest prematurely because of already accepted quadrant categorizations (the type that religions wants to pose as already confirmed hierarchies) then we can and will succumb too easily and too readily and too naively to the *Transcendental Temptation*. Or, to invoke Ken Wilber's pithy parlance, you cannot make a pre-rational and trans-rational fallacy distinction (distinguishing that which is truly within the five senses from that which is not) unless you have a deep and rich and nuanced understanding of all that is indeed pre-rational. How else can one determine that which is truly trans-rational?

A science of consciousness, therefore, must start with the brain. Science, in other words, can indeed point to that which is not physical because of its ultra-focused aim. Science can upend itself quite easily. The fact that it hasn't yet is why we remain so confident in its methods and its discoveries. But if in the future it comes across something that cannot be reduced to the four forces of the universe, we will be forced to reconsider. But what has happened in the past and what is still happening today is that we want to invoke transcendent explanations too quickly in order to salvage a sense of the numinous, forgetting in the process that even if all things are indeed material bits the mystery of all this (and here comes the pun) isn't lessened by one bit.

While I can appreciate that Davidson and others believe in a higher intelligence, we should not confuse our own faith with how the world works. We should also not provide false caricatures of how science is done in order to elevate our own theological biases.

Davidson in his desire to make a Consciousness Is All proposition succumbs to making a number of solipsistic errors. On page 147 and 148, for instance, he argues that we cannot "compare" or "tally" our experiences of the world with others. He writes, "I look at a chair, you look at a chair. What's the problem? The problem is that we have no way of comparing our two experiences to see if they tally. Perception is subjective."

While it is certainly true that perceptions are subjective, it doesn't then follow that we cannot correlate our experiences of a chair with others. We do it all the time. That is why we don't see two or three or more people sitting in the same chair, unless indulging in some sort of sophomoric prank. Moreover, neuroscience has made tremendous progress in developing ways for our subjective experiences to be understood and explained in more objective ways. Just recently New Scientist reported that, "Philip Low, of Stanford University in California, is testing a portable device called the iBrain to record Stephen Hawking's brain activity and identify what they mean."

I think the fundamental difficulty Davidson and perhaps other of a similar ilk have with science is that they have a mistaken understanding of the term matter and what it portends. Matter itself is not one thing and thus it isn't surprising that there is an almost infinite spectrum of possible variations, including different perceptual modes (from a dolphins using 3-D sonar or a bat using ultrasound or a dog extreme sensitivity to smells, to cite but three of Davidson's examples on page 148).

But Davidson doesn't want to explain these variations as the result of material processes but rather of a mind that he believes is "a separate and more subtle entity than both the brain and physical reality." This unnecessary dualism, of course, is richly ironic given the title of Davidson's book, *One Being One*.

It is here that Davidson's overall metaphysical agenda comes into full relief. He cannot imagine how matter could be responsible for consciousness and thus argues that the "the fundamental reality of all things is being or consciousness, not material substance."

But how does Davidson know this to be ontologically true? By his a priori reasoning skills? Thankfully issuing such grandiose statements as if by a divine fiat is not how science progresses. The very hubris he projects on science boomerangs back on Davidson when he makes such dogmatic remarks. Is it any wonder that skeptics will not find his rhetorical arguments convincing?

Davidson adds more to his mystic dogmatism when he asserts (without qualification) "The One Being's primary 'unit' of creation is not made of matter. His fundamental unit is a little being, the soul, made of His own 'substance'. In his own image--being or consciousness."

I can appreciate that such religious axioms might be expected for a catechism class for newly indoctrinated members of Radhasoami, but to pass it off in a book that pretends to be scientific is (and I will try to be polite here) a tad ridiculous.

Davidson believes that it is the mind that "creates the multiplicity of changing forms." He also points out that "most scientists don't like the idea of a 'ghost in the machine,' holy or otherwise." But he then proceeds to complain that science should not "always have the last word on everything."

But science never does claim to have the last word on anything. That is precisely why it is progressive and why technologies change and improve over time. Science is never at a standstill and is constantly questioning and correcting itself. Can the same be said of Radhasoami? When was the last time a mystic in such a tradition contradicted a previous master or a previous doctrine in the lineage? Radhasoami, and similar mystical schools, are the ones attempting to have the "last" word and Davidson repeatedly displays an unquestioning acceptance of shabd yoga mysticism over and above what science discovers. His book is not a reconciliation of science and religion. Rather, it is an unabashed apologetic to elevate a mystical worldview above all others and never deeply question his own religious tradition.

Davidson argues on page 123 that the "universe itself is supremely and surprisingly biofriendly." While those who are presently alive and kicking on this planet may think so, the facts tell us another story. Perhaps a laundry list of where life doesn't flourish will give us a better context in which to appreciate the not so "biofriendliness" of the cosmos at large. There is apparently no life (as we presently recognize it) on any of the billions of stars or countless black holes or innumerable planets throughout our own milky way, since they are in essence uninhabitable. Where then does life bubble forth? As far as we can decipher (given our limited resources at this moment) life occurs very rarely and then usually in fairly horrific conditions where eat or being eaten rules the day and night. Most of life that issues forth does so only after massive deaths of organisms that survive only briefly and even then under such dire circumstances that one can only wonder what anyone means by such a term as "friendly." In any case, the universe is filled with an astronomical amount of unusable and unlivable territory (statistically speaking, 99.9 percent or more of the cosmos is mostly not kind to life), which, of course, led one astute thinker to quip, "If there really is a God of this universe, one thing is certain: He is very fond of lots of empty parking spaces."

Clearly, we suffer from a myopic vision of how the universe operates when we only look our own survival as indicating something about how the everything is constructed. Viewed from a less anthropic purview, one can better appreciate Darwin's quote from Tennyson when he insightfully (and with an ear for a good pun) opined, "nature is read [red?] in tooth and claw."

A winner of a super lottery may reflect upon how fortunate she is to win such a large prize given the circumstances, and may think that it was "meant to be" given that any change in her routine would have led to a different outcome, since she allowed the computer to pick her number at random. Indeed, she might even think of every event up to that moment being special and interrelated since it led to such a stupendous result. However, what she doesn't think about are the millions of ticket buyers who lost and how they were not so fortunate. And how did she win? Simple answer: by chance. Understanding probability and how it applies in such circumstances gives us a much deeper insight than does our misguided belief in "it was our karma" type of thinking.

The same applies, even if we are resistant to admit it, to why our planet houses life and others do not. Or, to put this in sharper relief, if creating life forms was the purpose of the universe, the designer has surely done a terrible job given the overwhelming abundance of unused space and the torturous ways that even the simplest of creatures must endure to survive for barely seconds given the abundance of "eating" competition.

But Davidson sees none of that and instead claims on page 132 "that everything seems to be perfectly arranged for the existence and maintenance of life." Perfectly arranged? Apparently there must be a typo in his use of the word arranged, since a better choice would have been *deranged*.

Yet, it is in his chapter entitled "A Bunch of Old Fossils" near the end of the book where Davidson truly goes off the rails and indulges in exaggerations bordering on dishonesty. For reasons I cannot understand (to echo a similar refrain from Davidson about Christians and their take on creation), Davidson consistently misrepresents the current work being done on evolution and in so doing sideswipes a proper understanding of the revolutionary discoveries that have been made in the past fifty years. He even goes so far as to butcher a quote (without attribution) in his rush to make a sweeping (and inaccurate) allegation that "many" [his words] scientists have reached the conclusion that the "dynamic complexity of living organisms" cannot be reconciled with "underlying statistical probabilities." Of course, no names are mentioned, no references are provided, and his use of the quote "The probability that random chance created life is roughly the same as the probability that a hurricane could blow through a junkyard and create a jumbo jet," comes from Fred Hoyle and has now become known among evolutionary theorists as "Hoyle's Fallacy." Ian Musgrave's widely cited book [from talkorigins.org to Wikipedia, etc) *Lies, Damned Lies, Statistics and Probability of Abiogenesis Calculations* explains it thusly,

"These people, including Fred [Hoyle[, have committed one or more of the following errors. They calculate the probability of the formation of a "modern" protein, or even a complete bacterium with all "modern" proteins, by random events. This is not the abiogenesis theory at all. They assume that there is a fixed number of proteins, with fixed sequences for each protein, that are required for life. They calculate the probability of sequential trials, rather than simultaneous trials. They misunderstand what is meant by a probability calculation. They underestimate the number of functional enzymes/ribozymes present in a group of random sequences."

Davidson on page 157 even cites the widely criticized canard (best epitomized in Michael Behe's book, *Darwin's Black Box*) on page 157 that "The biochemistry of even the simplest microorganism is just too incredibly complex for it to have come into being by chance."

One wonders whether Davidson is even aware that this very notion of irreducible complexity has been severely criticized by molecular biologists who have gone out of their way to explain how evolution at such a level works.

As the *talkorigins.org* points out:

"Michael Behe's term 'irreducible complexity' is, to be frank, plainly silly — and here's why. 'Irreducible complexity' is a simple concept. According to Behe, a system is irreducibly complex if its function is lost when a part is removed. Behe believes that irreducibly complex systems cannot evolve by direct, gradual evolutionary mechanisms. However, standard genetic processes easily produce these structures. Nearly a century ago, these exact systems were predicted, described, and explained by the Nobel prize-winning geneticist H.J. Muller using evolutionary theory. Thus, as explained below, so-called 'irreducibly complex' structures are in fact evolvable and reducible. Behe gave irreducible complexity the wrong name. Behe claims that irreducibly complex systems cannot be produced directly by gradual evolution. But why not? Behe's reckoning goes like this:

(P1) Direct, gradual evolution proceeds only by stepwise addition of parts.

(P2) By definition, an irreducibly complex system lacking a part is nonfunctional.

(C) Therefore, all possible direct gradual evolutionary precursors to an irreducibly complex system must be nonfunctional.

Of course, Behe's argument is invalid since the first premise is false: gradual evolution can do much more than just add parts. For instance, evolution can also change or remove parts (pretty simple, eh?). In contrast, Behe's irreducible complexity is restricted to only reversing the addition of parts. This is why irreducible complexity cannot tell us anything useful about how a structure did or did not evolve.

With Behe's error now in hand, we immediately have the following embarrassingly facile solution to Behe's "irreducible" conundrum. Only two basic steps are needed to gradually evolve an irreducibly complex system from a functioning precursor:

Add a part. Make it necessary.

It's that simple. After these two steps, removing the part will kill the function, yet the system was produced directly and gradually from a simpler, functional precursor. And this is exactly what Behe alleges is impossible.

As a scientific explanation, the Mullerian two-step is extremely general and powerful, since it is independent of the biological specifics of the system in question. In fact, both steps can happen simultaneously, in a single event, even a single mutation. The function of the system can remain constant during the process or it can change. The steps can be functionally beneficial (adaptive) or not (neutral). We don't even need to invoke natural selection in the process — genetic drift or neutral evolution will do. The number of ways to add a part to a biological structure is virtually unlimited, as is the number of different ways to change a system so that a part becomes functionally essential. Plain, ordinary genetic processes can easily do both."

But Davidson appears uninterested in a real dialogue with science since it would point by point contravene each of his contentious and outdated pastiches. How else can he write (apparently with a clear conscience) such howlers, as "There is certainly no clear evidence that life and consciousness originated from material substance." To the contrary, there is overwhelming evidence that life and consciousness are indeed the result of material processes. It is so obvious, in point of fact, that one winces at Davidson's arrogance when making such an all encompassing and wrong-headed statement, particularly in light of the discovery of the structure of DNA by Watson and Crick (at Davidson's alma mater no less) at Cambridge University in 1953 and the astonishing breakthroughs in unraveling the genome of human beings and other organic creatures.

And adding to his apparent ignorance of the field, Davidson goes on to univocally state on page 160 that such an empirical study is "an entirely speculative theory, an unsupported extrapolation of the available data into the realm of scientific mythology."

It is at this point that any reasonable reader may reflect on why Davidson is being so dishonest and disingenuous about the vast array of evidence that the varying sciences have marshaled in support of evolution by natural selection. The answer, sadly enough, is too obvious: Davidson is a religious devotee who believes (apparently absolutely and apparently quite dogmatically) in the metaphysical theology of Radhasoami Beas and anything which upends his cherished belief system is open game, even if it means being duplicitous in his presentations of the "other" side.

One Being One isn't genuinely interested in a science-religion dialogue, but rather is bent on being a polemic against all that which contravenes Davidson's and Radhasoami's somewhat medieval metaphysical schema. He even impugns the integrity of evolutionists alleging that "such an interest [in that which would contradict the current Darwinian paradigm] is likely to be seriously bad for their career prospects."

Yet, Davidson doesn't seem to realize that if a scientist (following the evidence and not merely mystical dogmatism) were to discover something that upended some or all of Darwin's theory of evolution by natural selection, he or she would be heralded as a pioneer in the field. Scientific awards, lest we forget, are given to those who discover things we don't know. The Nobel Prize in physics, chemistry, or medicine, for instance, is not given to those who merely confirm what we already know, but to those who make a singular or unique contribution.

In assailing Darwinian evolution (but without mustering convincing evidence), Davidson on page 164 juxtaposes molecular biology with his own mystical theology, one where "behind all the physical processes lie the patterning processes of the formative mind, projecting subtle inner mental patterns into physical reality. This is all part of what some folk call the law of karma."

Does Davidson give us even a scant of evidence for this theory? No. So how does he know that his subtle mind theory is correct, especially in contradistinction to the plethora of evidences provided by the hard sciences? The answer again is a revealing one: because his religion says so and because he believes in the veridicality of meditational experiences. Yet, what is so alarming, particularly in a book that wants to be taken seriously, is that Davidson doesn't once find even the tiniest fault or crack in his own theological superstructure. In this sense, he is no different than fundamentalist Christians or Muslims who take their respective holy books as inerrant.

Davidson, not the science and scientists he so mistakenly criticizes, is the one proffering a dogmatic mythology, especially when he can make such axiomatic statements (as on pages 164-165) as "In the divine process, consciousness gives rise to matter, not the reverse. Mind and spirit give rise to bodies. That's a fundamental principle [my emphasis].

Davidson isn't providing us with a "science" of the soul, but a well-worn dogma disguised in pseudo-scientific language that pretends to be something that it is not. Quite frankly, *One Being One* is a dishonest book written not so much to bridge the gap between science and religion, but to not so subtly proselytize for Radhasoami's version of intelligent design. Even most young earth creationists are more upfront about their agendas than Davidson is in his tome.

What is it about matter than makes erstwhile monists succumb to the very dualism they wish to transcend? Davidson harps throughout his text about the One Being One, but repeatedly resorts to a series of outdated models, based more on Hindu philosophy than anything offered by neuroscience, which display a dualistic chain of being, more similar to Gnosticism than to absolute monism.

The penultimate chapter, "Present in the Presence," of Davidson's book, betrays its real thesis and it is here that he gives up all pretense at being objective and simply tries to persuade his reader to convert to his religious way of thinking, even as he still tries to soft pedal his spiritual counseling as a some sort of universal dictum without cultural restrictions. On page 177 Davidson preaches, "The divine Beloved is our guide, drawing us ever one. Our effort is simply a response to his call. 'If we taken one step towards Him, He takes a hundred steps toward us.' And He is the one who makes us take that one step. His grace is inestimable, His love is incalculable."

While such proselytizing and preaching may be appropriate for a religious catechism book, it seem entirely inappropriate and even knavish in a text that wishes to honestly present a "reconciliatory perspective" between science and religion. Even the imprint "Science of the Soul Research Centre", which is now how the Radhasoami Beas religion wants to advertise itself in the West, seems contrived and designed to give the impression of some sort of objective science enterprise which it is clearly not.

What kind of open research is Radhasoami Beas truly engaged in, except the reiteration of their own philosophy and the regurgitation of their own mystical practice--one that is guided by a supreme leader who the majority of followers believe to be god in human form? There is something deeply unseemly in this approach, particularly when Radhasoami is anything but a science in the general sense of that term and when using such an advertising ploy never engages in a critical analysis of its own belief system. Science progresses precisely because it allows for corrections and is willing to change its core ideas over time when better and more accurate theories emerge. Has Radhasoami Beas done anything akin to this? No. Perhaps it would be better for all concerned if Gurinder Singh and Radhasoami Satsang Beas stopped the charade of pretending to be a science and came to grips with the fact that it is a mystically oriented religion like many others around the world.

Perhaps if John Davidson dropped the ruse that his book was about science and more about his own theological belief system his readers would be better informed about his real purposes and could accept it for what it ultimately turns out to be: a fundamentalist religious tract for Radhasoami Satsang Beas.

4 | The Feynman Imperative

Don Salmon wrote some very interesting notes and pointed questions on our recent review article, *Mysticism's Version of Intelligent Design*, on the commentary page of Brian Hines' blog *The Church for the Churchless*. The following is my rejoinder.

Don Salmon: I'm always intrigued at how much people believe the catechism of scientists. Yes of course, scientists like Richard Feynman no doubt believe what they say when they point out that experiments are not determined by belief. Might I point out that Feynman was a physicist who by virtue of that statement shows he knows little or nothing of the vast research in cognitive science showing how profoundly ignorant we are of the extent to which our beliefs shape our conscious "rational" thinking (and as an interesting aside, have you noticed that the new spate of books in the field of neuroscience purporting to show us how easily our beliefs overrule our rational side, almost all share a physicalist perspective, and almost to an author direct all their attention at showing how any body with a remotely non-physicalist belief is clearly irrational, and it never seems to occur to them that it might just be remotely possible to turn their attention to their own – may I say it – "beliefs" and then they might find out that this rational, belief-proof evidence they make such high and mighty and appealing noble claims about is no more impervious to bias, prejudice and dogmatic blindness than the most intransigent medieval churchman?)

David Lane replies: Richard Feynman, contrary to your assumption, was not naive about the manifold ways that belief can influence the way we go about doing science, including how experiments are designed. But the key point I was making, which I should have emphasized more clearly, was that science makes progress because it follows the trail of evidence. One of the best ways of getting such evidence, of course, is by experimentation. This doesn't mean that experiments are value free or devoid of human bias. That is why reproducibility and the ability to make unexpected predictions is one of the cornerstones in science.

In other words, what a scientist wants to discover is whether a particular phenomena or observation will hold true regardless of his or her beliefs or biases. It may be one thing for a scientist to claim that he has discovered cold fusion, for instance, but the linchpin is whether others (who may not hold such a belief, but who are willing to "test" it) can reproduce those results. If we discover that it cannot, then naturally our skepticism grows. As Feynman explained,

"In summary, the idea is to try to give all of the information to help others to judge the value of your contribution; not just the information that leads to judgment in one particular direction or another."

Feynman was also acutely aware of how easy it was to be deceived by our own desires and beliefs and prejudices. In an illuminating commencement speech given at CAL TECH in 1974, later known more popularly as Cargo Cult Science, he said, "The first principle is that you must not fool yourself--and you are the easiest person to fool.

Don Salmon: By the way, have you actually done any research or hung out with people doing research? A few weeks of conversation with researchers would, I think, quickly reveal the extent to which bias and tightly held beliefs pervade the scientific endeavor.

David Lane replies: Yes, that is one of the requirements one has to do when securing a Ph.D. and a couple of M.A.'s. And, yes, it is little wonder that biases and "tightly" held beliefs pervade those doing science. But, again, what makes science work and progress is that it is predicated not upon what we desire or wish, but rather competing our guesses, hunches, and theories by allowing them to be tested in real world situations and against other hypotheses and other viewpoints. In this remarkable cauldron we uncover those ideas that survive what Karl Popper famously called "falisfiablity."

Any ism can do the same (science is, after all, a human enterprise and open to anyone, anytime, anywhere), but to qualify as scientific it must do that most remarkable of things: it must be willing to be disproven, to be corrected, to admit to error. Let me illustrate this from an incident that happened to me when I bought the first Apple iPad back in 2010 and brought it to the computer technical support team at my college. I wrongly assumed that they would be mesmerized by the new gadget and I was expecting some "oohs and aahs." That is not what I got. Instead, almost from the moment they placed their hands on the touch screen and began swiping this or that app, they voiced complaints. "It doesn't have flash? That sucks." "No camera?" "No SD slots?" And so on. I was taken aback and thought to myself that these techies don't appreciate a good product when it stares them in the face. But then later that night I smiled and realized that they were right to be so critical. Why? Because seeing the flaws or the mistakes or the weaknesses in something (and imagining better ways to do it) is precisely how such tablets will be improved in the future. And so they have, as anyone with an iPad 2 or an iPad 3 will attest.

Science works (and advances) because it is rooted in trying to spot weaknesses or flaws or insufficiencies. It is in a way conscious map-making, where the would-be cartographer is constantly surveying possible gaps in the best up to date topographies.

Undoubtedly there will be those who claim to be scientists who will try to cheat or pass off their results as scientific. But the great thing about science (and why it has had such a successful run) is that it is constantly open to correction. Nothing is sacred, not even the most cherished of ideas. No authority is beyond questioning. Einstein or Darwin or Hawking or Witten or whomever can be and has been wrong.

As Feynman explains, "The scientist has a lot of experience with ignorance and doubt and uncertainty, and this experience is of very great importance, I think. When a scientist doesn't know the answer to a problem, he is ignorant. When he has a hunch as to what the result is, he is uncertain. And when he is pretty darned sure of what the result is going to be, he is in some doubt. We have found it of paramount importance that in order to progress we must recognize the ignorance and leave room for doubt. Scientific knowledge is a body of statements of varying degrees of certainty -- some most unsure, some nearly sure, none absolutely certain." One doesn't have to believe in gravity, for instance, to see and acknowledge its effects.

Don Salmon: A complete red herring; or perhaps that's the wrong phrase. The specific phrase of yours I'm challenging as involving a belief system is the idea that science has the capacity to judge the source of consciousness (an obvious impossibility when you consider that scientists as of yet do not accept any methodology by which one could even detect the presence of consciousness – I'm including the flimsy attempts among qualitative researchers to attempt some form of introspection, which as currently practiced also does not provide any means of detecting consciousness from a scientifically acceptable perspective). As far as gravity, the only aspect of it relevant to your phrase is whether "gravity" is independent of consciousness. And when I say "gravity" – of course we realize we're not talking about anything we have direct empirical evidence of - the quantum physicist's phrase (I think it was Eddington but I'm not sure) "something is doing something to something that we know almost nothing about" – yes, we feel something, and our instruments (which also are only known to us as forms in awareness) respond and we hypothesize the activity of something to which we give the abstract name "Gravity" – this has nothing to do with whether there is some force that exists entirely independent of consciousness.

David Lane replies: I find it a bit odd and perhaps a bit telling that you seem almost to have an a priori resistant to the notion that consciousness has a physical basis. Are you sure that your own belief system isn't blinding you from at least the possibility that self-reflective awareness is grounded within a physical substratum? You, and this is not without irony, make a grandiose and absolute claim when you write that you challenge "... the idea that science has the capacity to judge the source of consciousness (an obvious impossibility when you consider that scientists as of yet do not accept any methodology by which one could even detect the presence of consciousness) [my bracket closure]. . ." Impossible?

I realize that you have already shown your guiding metaphysic when you cited a passage from Krishna Prem, "It should be clear from introspective meditation that all forms are sustained in consciousness, and that, apart from consciousness, we know nothing and can know nothing of forms. It is in fact meaningless to talk of forms as existing apart from consciousness The objects supposed by some to exist behind the forms are mere mental constructs devised for dealing with experience in practice. No one knows them, no one can ever know them; to believe in their existence is a pure… act of faith."

While I can appreciate this Consciousness is All (or CIL) proposition, since we are indubitably stuck to our own awareness, which invariably filters what we know about the universe at large and within, it doesn't follow that such reflective awareness cannot be the result of underlying physical properties. We should be cautious not to confuse our state of awareness with how that state came into being.

V.S. Ramachandran studies on phantom limb pain are instructive here, unless we want to fall into an infinite labyrinth of solipsism. As "Rama" (as he is known among his colleagues and students) wrote, "With the advent of non-invasive imaging techniques such as MEG (magnetoencephalogram) and functional MRI, topographical reorganization can also be demonstrated in humans, so that it is now possible track perceptual changes and changes in cortical topography in individual patients. We suggest, therefore, that these patients provide a valuable opportunity not only for exploring neural plasticity in the adult human brain but also for understanding the relationship between the activity of sensory neurons and conscious experience."

Now from those suffering from phantom limb pain (and not conversant with the neural mapping which causes it), they feel as if they still have an arm or a leg and that the pain is located in one or more of their now amputated extensions. Their own conscious experience doesn't reveal that the origination of such pain is nowhere in an extended limb but is rather part and parcel of their brain.

Once this is pointed out to them (i.e., once they learn to cognitively bypass their own conscious presumptions), they can be quite successful in either eliminating or greatly reducing the pain. Thus, we should be quite cautious about conflating our experience of consciousness with how such self-awareness originates.

Perhaps one of the key issues you bring to bear is when you write of the impossibility "that scientists as of yet do not accept any methodology by which one could even detect the presence of consciousness."

I am not quite sure what you mean here by "detection" but perhaps it is similar to Sam Harris' dilemma where he writes, "The problem, however, is that no evidence for consciousness exists in the physical world. Physical events are simply mute as to whether it is 'like something' to be what they are. The only thing in this universe that attests to the existence of consciousness is consciousness itself; the only clue to subjectivity, as such, is subjectivity. Absolutely nothing about a brain, when surveyed as a physical system, suggests that it is a locus of experience. Were we not already brimming with consciousness ourselves, we would find no evidence of it in the physical universe—nor would we have any notion of the many experiential states that it gives rise to. The painfulness of pain, for instance, puts in an appearance only in consciousness. And no description of C-fibers or pain-avoiding behavior will bring the subjective reality into view."

I have written an extended article on Harris' conundrum ["Inside Outside: Sam Harris' Dilemma"] and illustrated his concern by employing a Mobius strip as a fitting metaphor. In this regard I don't think we have to disregard consciousness (Searle's 1st person) and only opt for a purely objective (Searle's 3rd person) perspective. I don't see why it has to be one versus the other. And such notable scientists as Gerald Edelman, Nobel laureate and pioneering thinker in neural Darwinism, and others have long argued that both perspectives must be incorporated.

But in taking this approach, I also don't see why we have to be resistant to exploring the physical basis of awareness. Indeed, I have long argued that such reductionism is altogether progressive, even if one ends up with a purely non-material explanation. In order to accomplish this aim one doesn't have to separate the experience of awareness from its empirical mooring.

I am curious as to why you are under the impression that science has to do away with consciousness in order to understand it. Science is well aware, certainly after the breakthroughs in quantum mechanics of the 1920s and the mathematical revelations of Kurt Gödel, Alonzo Church, and Alan Turing of the 1930s, that one cannot naively escape the observer component in studying nature.

You write, "[when] . . .we hypothesize the activity of something to which we give the abstract name 'Gravity' – this has nothing to do with whether there is some force that exists entirely independent of consciousness."

But therein I think is why we may be talking at cross-purposes. Arguing that consciousness has a material basis doesn't then mean by extension that we have to make an ontological claim such as gravity or any other force is "entirely" independent of consciousness. That is not necessary, just as when physicists making graphene don't have to invalidate Heisenberg's principle of uncertainty when they isolate an atomic plane of graphite. Moreover, I am not arguing for a naive dualism of matter and awareness as separate entities, since I think the real blockade is mostly a linguistic confusion over the word matter as indicating something grey, flat, or one-dimensional. When in point of fact and in point of experimentation, it is anything but.

It was for this very reason that my wife, Andrea, and I presented an extended film presentation via video hook-up (along with an extended illustrated publication) to the International Conference on Spiritual and Consciousness at the Dayal Bagh Educational Institute in Agra, India, in 2010 explaining that religion in general, and more mystically oriented paths in particular, need to change and update their understanding of what the term matter actually means. This is why we wrote,

"Therefore, a very strong argument can be made that the real problem with materialism (the idea everything that arises is nothing but permutations of matter) isn't that it is the exact opposite of spirit or that it somehow diminishes human consciousness, but rather that we do not properly appreciate what the word actually means and what it entails. To say something is 'just' matter is akin to say something is 'just' light (which matter, by the way, also contains). Even when intertheoretic reductionisms hold true there is no "just" about it, since the very phenomena under inspection doesn't lose its mystery by being contextually or algorithmically comprehended. If someone says that the Atlantic is merely H20, the ocean and all its magnificence isn't lessened by such molecular equations. The trouble isn't with matter or our tendency to ground all properties to it, but rather that we are assuming that matter is one thing when it is completely the opposite of that."

The same holds true with electromagnetism. Nathaniel Branden provided a wonderful insight into the weaknesses of some of Ken Wilber's analogies and I think the same applies here.

Don Salmon: What was said about gravity equally applies to electromagnetism. I assume, being a physicalist, you have the assumption that electromagnetism exists independently of any kind of consciousness whatsoever. Maybe it does. Whether it does or not is not my interest here. I'm not saying you're "wrong". I'm only pointing out that in your comment there is a non-empirical assumption which cannot in any way – given currently acceptable scientific methodology – even be put to an experiment much less be proven or disproven. It is utterly meaningless in the current scientific framework.

David Lane replies: Your last paragraph nicely summarizes a fundamental misinterpretation you have about my views on consciousness as being physically based. You assume that I believe that electromagnetism "exists independently of any kind of consciousness whatsoever." I have never said such a thing nor does any physicalist have to hold such a supposition, particularly in light of the statistical and probabilistic basis of fundamental physics where our experimentations (or intrusions) alter what is observed.

No, my position is that consciousness is simply a different aspect of matter. Rearrange atoms in such a way and you get the elements in the periodic table--from hydrogen to iron to plutonium, etc. Rearrange those same atoms in particularly complex ways and you get organisms with the ability for virtual simulation and self navigation/reflection. We don't have to opt for Cartesian style dualism or opt for silly outdated and inaccurate definitions of matter.

One may object and say that we cannot study anything without our attendant awareness and thus we are stuck (whether we like it or not) with an all pervading consciousness lighting up everything that we see and pontificate upon. Yes, but saying this doesn't then mean that consciousness cannot be physical. It simply means we have to reorient ourselves to a deeper understanding of the wondrous plasticity and multi-dimensionality of what we mean by matter in the first place. As for Sir Arthur Eddington's quote, it apparently refers not to gravity (but, like you, I can readily see how it could), but to his observation on Heisenberg's principle of uncertainty in 1927.

Don Salmon: Here's something from John Searle that is relevant: "How is it that so many philosophers and cognitive scientists can say so many things that, to me at least, seem obviously false?... I believe one of the unstated assumptions behind the current batch of views is that they represent the only scientifically acceptable alternatives to the anti-scientism that went with traditional dualism, the belief in the immortality of the soul, spiritualism, and so on. Acceptance of the current views is motivated not so much by an independent conviction of their truth as by a terror of what are apparently the only alternatives. That is, the choice we are tacitly presented with is between a "scientific" approach, as represented by one or another of the current versions of "materialism," and an "unscientific" approach, as represented by Cartesianism or some other traditional religious conception of the mind. Sunday, July 29, 2012"

45

David Lane replies: I have long enjoyed Searle's books and lectures, especially his acerbic style, but a closer reading of Searle reveals that he too thinks that consciousness has a material basis. He is very clear about it, especially when he writes, "We know consciousness happens and we know the brain does it."

The question for Searle, therefore, is not if consciousness is materially produced, but how a complicated and interwoven three pounds of glorious meat does it. Searle isn't objecting to the physics of awareness, as such, but to the proposed models and maps of how the brain produces self reflective awareness, particularly those which have taken a purely computational approach.

There is nothing "metaphysical" about using anesthesia in a hospital to perform surgery on a patient, thereby rendering his "waking" state consciousness dormant. There is nothing "metaphysical" when that same patient is kept under sedation by such "physical" chemicals as cyclopropane.

Don Salmon: I'm assuming that you're taking "anesthesia" to be a purely non-mental or non-conscious phenomenon; similarly, you apparently have the same assumption about cyclopropane. When the only way you or anybody could possibly know about "anesthesia" (do you mean the sensory phenomenon or something that is measured by certain instruments) or any chemical or any instruments which provide information about chemicals is as a form in awareness, then your description of something as "physical" – along with the assumption that "physical" means independent of consciousness – if not "metaphysical" – then in more simple psychological terms, is a virtually non provable assumption..... imagine trying to prove something is consciousness-independent – as soon as you think you've ascertained it, you've made it an object of consciousness. And what could the word "object" mean independent of subjectivity?

David Lane replies: Once again you mistakenly assume that I think matter and consciousness are two different properties, especially when you persist in stating that "the assumption that physical means independent of consciousness." Again, I have never stated such a dualism, since I don't think it is at all necessary to do. Just as John Searle has pointed out that water has many aspects (including the property of wetness) it doesn't then follow that we have to ontologically separate such categories, even if we may distinguish their varying emergent properties.

Once again, I think the confusion is systemic and stems from the notion that we cannot get out of our own awareness to see how such a process could arise without inevitably seeing it through the lens of our own reflections on the subject. Yet, that is why it is important to open up our understandings to others who may have a different view or perspective. Science focuses us not to fall prey to our own solipsism. And, as such, allows us to communicate our subjective offerings as open sources so that others (with similar or varying operating systems) can appraise or adjudicate them in larger arenas and larger testing.

Studying consciousness scientifically doesn't necessitate eliminating the 1st person narrative and only allowing a third person description. To the contrary both can go hand in hand.

No, I think confusing science's practical tools with metaphysics is to entirely misunderstand the very nature of how and why science progresses.

Don Salmon: Once again (leaving aside the term "metaphysical" for now) we have an idealized view of science which bears little or no resemblance to the incredibly messy, emotional, prejudice filled world of real research. I've seen scientists defend this by saying, "well yes that's how research really works, but still, we hold that before ourselves as an ideal." Yes, and if you believe the stated ideals of various professions, then there is no such thing as police corruption, all corporate CEOS are saints, and all politicians – well, we won't go there...

David Lane replies: I don't think we have an idealized view of science at all. Rather, it has been pretty clear to anyone who has either engaged in science or who is familiar with books on the sociology and history of science that it is indeed a thoroughly human affair and, as you point out, quite a messy one. Generally speaking, no one is claiming otherwise. Yet, that ironically is its greatest strength since science cannot hold out absolute truths and must by its very nature allow competing ideas and competing theories to be played out. Scientists are invariably finding faults with research design, statistical analyses, experimental protocol, and studies which make hasty inferences and conclusions. That is why science is tentative and is always looking, indeed encouraging, others to test and retest observations and findings, since there can always be a missing part of the puzzle.

My wife, Andrea, worked as a research assistant to V.S. Ramchandran, the famed neuroscientist, on visual perception at UCSD for a year or so and she did original studies where oftentimes the results were mixed and didn't necessarily support one's own pet hypotheses or fit into the prevailing paradigm. But that is to be expected.

As Richard Feynman rightly pointed out, "We've learned from experience that the truth will come out. Other experimenters will repeat your experiment and find out whether you were wrong or right. Nature's phenomena will agree or they'll disagree with your theory. And, although you may gain some temporary fame and excitement, you will not gain a good reputation as a scientist if you haven't tried to be very careful in this kind of work. And it's this type of integrity, this kind of care not to fool yourself, that is missing to a large extent in much of the research in cargo cult science."

It is for that reason that I think exploring the very material basis of consciousness is the key and why, lest we forget, we want our doctors trained in how trimethylene actually works in a real physical body and not going off ruminating about how science is merely a belief system or entirely metaphysical.

Don Salmon: What do you mean by "material" or "a real physical body"? Whatever it is, underlying it is the assumption that what is "material" or "physical" is consciousness-independent.

David Lane replies: No, I don't think your logical syllogism holds. Positing the physical basis of consciousness doesn't mean by implication that "material is consciousness independent."

Those who explore the very heart of quantum electrodynamics or quantum chromodynamics and who may philosophically agree with Niels Bohr and others in their Copenhagen interpretation of subatomic physics ("It is wrong to think that the task of physics is to find out how Nature is. Physics concerns what we say about Nature.") focus on what they can uncover and have made tremendous progress in their pursuits. There is absolutely no reason why neuroscientists exploring the brain-consciousness connection cannot do the same. It is an unnecessary dualism that you are invoking and one that is not required to explore the neuronal basis of awareness.

Don Salmon: Again you may be right, but what materialists and physicalists usually believe is that "that's just the way things are" and it has nothing to do about ruminating about belief systems or metaphysics. I'm just point out that there's a startling amount of ruminating, assumptions and hidden beliefs in the use of the term "material" – particularly, "the very material basis of consciousness" – does that material basis have any association with consciousness "from the beginning" ("beginning' also being a word fraught with difficult unprovable assumptions – and please don't assume I'm advocating phenomenalism or skepticism; remember, I'm not advocating any position – not nondualism either – though that's not really a position, at least not in Nagarjuna's hands)

David Lane replies: Well, I am not altogether sure what you are driving at, since your sweeping generalizations about what materialists and physicalists "usually believing that's just the way things are" doesn't resonate with me nor does it with the scientists I know. And, interestingly, one of the reasons there can be progress in science is because researchers do hold certain positions and certain assumptions, but then they are required to "hang those out to dry" so that others may see where their particular ideas hold true and where they do not. Science only works to the degree that it can withstand an onslaught of competition, where eventually (by trial and by error) the best explanations--even if only tentatively held--hold court until another and better theory emerges. Newton's understanding of gravity was quite a breakthrough for its time, but Einstein's theory was more comprehensive and explained hitherto inexplicable problems that Newtonian physics could not. Such is science and how it works over time.

David Lane: Or, to give another example, planes fly and have been improved over the last 100 years not because science is a metaphysical system of beliefs, but because we test, and test again, very physical objects in very physical arenas.

Don Salmon: I don't think you're actually meaning to do this, but I find in these conversations with physicalists there's often a point where they say "well, where has your metaphysics (or epistemology or lucid dream experiments or whatever) gotten us in 3000 or so years; "science" has given us real useful things like flying planes and bombs and eyeglasses and cars; what do you want to give all that up?" That's often the conclusion, that I'm being "anti-science".

David Lane replies: Yes, I really do mean this. Science works because it gets tested in real world situations. I don't know if you are anti-science, but you do seem resistant to a physical explanation of consciousness and I think you have prematurely assumed far too much about what science can and cannot accomplish by closely examinng the physical parameters about how and why self-reflective awareness can arise when a certain bio-chemical complexity is reached.

Don Salmon: So much has the word "science" become associated with the unprovable, nonempirical belief system of physicalism that to challenge it is taken to be a challenge to science itself. Rather, I'd like to think I'm defending science against dogma, carrying on in the tradition of William James, who wrote more than 100 years ago: "Science taken in its essence should stand only for a method and not for any special beliefs, yet as habitually taken by its votaries, science has come to be identified with a certain fixed general belief, the belief that the deeper order of nature is mechanical exclusively, and that non-mechanical categories are irrational ways of conceiving and explaining even such a thing as human life."

David Lane replies: This seems like a classic straw man argument since I am not (nor are the cognitive scientists I know) defending scientism, which really isn't science at all, but rather dogmatism dressed up in cynical garb. You claim that "science [has] become associated with the unprovable, nonempirical belief system of physicalism that to challenge it is taken to be a challenge to science itself." Yet, you provide no specific examples with which to your hang your very questionable claim. What has any of this to do with the pioneering work being done in neuroscience and its interface with consciousness? My argument is a very simple one: let's explore the physics of awareness first and see where it leads us. If my hearing, seeing, smelling, tasting, and touching has a physical basis (e.g., just think of how successful eye surgery has become in the last two decades), is it really that much of a stretch to think that my "self awareness" is also grounded within my neural-body complex? I think not, and therefore instead of getting caught in endless philosophical debates on the subject, I suggest it is wise to take a more practical approach and let scientists in different fields (from neuroscience to mathematics) be encouraged in their efforts. Consciousness studies is still in its early days and I think we will surprised in the decades to come to see how much progress will be made on the purely physicalist front.

In conclusion, I don't think our humanity is lessened if we discover that you and I are the product of material complexity. Matter is as mystical (in the wondrous sense of that term) as anything posited in our religious scriptures. And, therefore, I think our resistance to empirical explanations is based on a deep misunderstanding of what the term matter actually defines and what it signifies.

Richard Feynman gave a beautiful talk in 1966 in New York City to the National Science Teachers Association which illustrated the beauty of doing science,

"The world looks so different after learning science. For example, trees are made of air, primarily. When they are burned, they go back to air, and in the flaming heat is released the flaming heat of the sun which was bound in to convert the air into tree, and in the ash is the small remnant of the part which did not come from air that came from the solid earth, instead. These are beautiful things, and the content of science is wonderfully full of them. They are very inspiring, and they can be used to inspire others. Another of the qualities of science is that it teaches the value of rational thought as well as the importance of freedom of thought; the positive results that come from doubting that the lessons are all true. You must here distinguish--especially in teaching--the science from the forms or procedures that are sometimes used in developing science. It is easy to say, "We write, experiment, and observe, and do this or that." You can copy that form exactly. But great religions are dissipated by following form without remembering the direct content of the teaching of the great leaders. In the same way, it is possible to follow form and call it science, but that is pseudo-science. In this way, we all suffer from the kind of tyranny we have today in the many institutions that have come under the influence of pseudoscientific advisers."

Science isn't a thing or merely a body of facts. It is a process, a human process no doubt, but one which works best only when we test our own pet guesses and hunches and see how well they hold up against other ideas and other hypotheses. This, in other words, is the Feynman Imperative: our willingness to be wrong.

As Richard Feynman very wisely explained, "I can live with doubt and uncertainty and not knowing. I think it is much more interesting to live not knowing than to have answers that might be wrong. If we will only allow that, as we progress, we remain unsure, we will leave opportunities for alternatives. We will not become enthusiastic for the fact, the knowledge, the absolute truth of the day, but remain always uncertain ... In order to make progress, one must leave the door to the unknown ajar."

5 | *The Practicality of Science*

I appreciate Don Salmon's openness for dialogue and for his willingness to take responsibility for any unclearness in his presentation.

In light of this, I fear that I don't fully understand why Salmon would then entitle his rejoinder "The Feynman Delusion" since there is nothing in his essay that illustrates where, when, and how Richard Feynman is delusional.

Indeed, I think the real delusion is in Salmon's persistence in repeating his straw man argument when he writes (yet again) the following,

"There is not a single scientific experiment that could ever be done (at least, not within currently accepted bounds of scientific methodology) to determine the absolute existence of some material or physical "stuff"/object/thing/anti-matter or whatever you want to call it, existing absolutely independent of lived experience."

Yet, nowhere in my previous essay have I argued for such an agenda. Rather, I wrote this, "Arguing that consciousness has a material basis doesn't then mean by extension that we have to make an ontological claim such as gravity or any other force is entirely independent of consciousness. That is not necessary, just as when physicists making graphene don't have to invalidate Heisenberg's principle of uncertainty when they isolate an atomic plane of graphite. Moreover, I am not arguing for a naive dualism of matter and awareness as separate entities, since I think the real blockade is mostly a linguistic confusion over the word matter as indicating something grey, flat, or one-dimensional. When in point of fact and in point of experimentation, it is anything but."

Perhaps it would be best to focus on how science actually works versus getting caught in an intractable (and, I would suggest, unnecessary) metaphysical debate over "physicalism."

Let me see if I can illustrate this better by a couple of examples. I have an old 1977 36 foot Roughwater boat called appropriately enough "The Phantom." Invariably, something always goes wrong each summer with the boat on our trips to Catalina Island. Usually it has something to do with the generator. On one particular voyage to the backside of the island to explore surf spots, I noticed that I couldn't get the generator to turn on. This was a real bummer since without the generator I couldn't get the onboard refrigerator to work nor could I charge accessory items, such as our portable GPS and our iPhones and iPads.

How to solve the problem? Science confronts difficulties like this all the time and how one proceeds tell us much about how science actually works (versus our idealized and abstract notions of it). As Feynman rightly points out in his now famous lecture on how to do science, we first "take a guess" about what we think is happening. We can, of course, makes all sorts of conjectures about why, in our case, the generator is not working. We could, even, if we were so inclined, make metaphysical guesses, such as there are evil astral goblins ruining the solenoid or that we don't have enough love and faith in the rubber belts. Eventually, however, no matter what type of guess we make, we have to conduct an experiment or "test" our hunches in time and space. We have to see, in other words, how well our guess will help us solve the problem. And in this sieving process of competing ideas, we discover that some hypotheses are better than others in getting the generator to work again.

In my situation, it turned out that the battery that kickstarted my generator was dead and had to be replaced. It was a simple, even obvious, solution. All my other guesses turned out to be less than sufficient.

A similar example can be drawn from oceanography where wave prediction has been fraught with difficulties for centuries for sailors and fishermen who did not have access to more complete models of how ocean swells are generated, sometimes thousands of miles away by intense winds. In years past, fishermen would imagine all sorts of reasons why waves appeared as they did, but most of such imaginings were too bounded by superstition and folklore to be properly testable.

Today, with the advent of much more sophisticated weather maps and detailed records of how hurricanes and other wind powered storms behave, wave prediction has become both a precise science and big business for the surf industry, as witnessed by the tremendous success of Sean Collins' company, *Surfline*, which provides predictive models around the world for professional and amateur surfer in countries from Australia to France to Fiji.

Practical science isn't necessarily concerned with ultimate philosophical issues, but with getting a particular problem resolved by testing out varying and competing hypotheses.

This is why Richard Feynman rightly points out that science is a process of discovery and that the ultimate progressive tool of any scientific endeavor is to see how well it explains (and predicts and resolves) any particular issue, problem, or mystery.

I therefore do not understand why Don Salmon continues to create a boogie man about science when he writes, ". . . not to say that I'm trying to prove that such independent matter or physical 'stuff' doesn't exist; that's not my intention—only to point out that such a belief is a nonscientific, nonempirical, unfalsifiable assumption."

Science is not predicated upon an already agreed set of beliefs (if it were, we wouldn't have the progressive technologies we see today). No, it is a practical affair where results (not ideologies) hold sway. If there really were gremlins that played havoc on generators and we had ample evidence of it, then science (as a process, not as a dogma) would naturally follow up on that line of reasoning.

Thus, the very reason science has tended to focus on the physical substratum of varying phenomena is because it has turned out to be a more fruitful line of inquiry. If it could be shown that King Neptune actually made waves off the coast of New Zealand, then that would be the working model of oceanographers.

But that hasn't been the case and thus Greek and Roman god conjectures haven't become the mainstay of our environmental textbooks.

I find it both disingenuous and quite odd that Don Salmon can indulge in such a false and misleading caricature of how science operates, especially related to the brain and self-reflective awareness, when he writes, "The physicalists have been telling us for more than a century and a half that it is a complete mystery how the stimuli from the 'outside world' set our optical, auditory and other nerves in motion and then—suddenly!—across some kind of mysterious 'gap'—experience takes place. But oh, according to the catechism of promissory materialism, by gum, they're going to find the answer."

Huh? First, before making such sweeping generalizations it might be wise to be laser specific about what "physicalists" one is referencing, since bundling all empirical scientists in one unified camp is not only completely misleading it is dishonest. Second, not every neuroscientist thinks it is a "a complete mystery" about how outside stimuli engenders optical and auditory experiences. To the contrary, there have been remarkable studies on precisely these issues and some have even led to a major breakthrough in restoring vision to erstwhile congenitally blind children.

"'Children who were treated with gene therapy are now able to walk and play just like any normally sighted child," said co-first author Albert M. Maguire, MD, an associate professor of Ophthalmology at Penn and a physician at Children's Hospital. 'They can also carry out classroom activities without visual aids.' Maguire and Bennett have been researching inherited retinal degenerations for nearly 20 years. Leber's congenital amaurosis (LCA), the target of this current study, is a group of inherited blinding diseases that damages light receptors in the retina. It usually begins stealing sight in early childhood and causes total blindness during a patient's twenties or thirties. Currently, there is no treatment for LCA. Children able to see after a single shot of gene therapy. For children and adults in the study, functional improvements in vision followed single injections of genes that produced proteins to make light receptors work in their retinas. Walking along a dimly lit, simulated street route, the children were able to negotiate barriers they bumped into before the surgery. Another child, who since birth, could only see light and shadows, stared into his father's face and said he could see the color of his eyes. Later they played soccer together.'"

By centering on the very physical processes of seeing scientists found a procedure in which to improve retinal vision in both children and adults. And so have similar breakthroughs occurred in hearing and smelling. I know the latter from my own personal experience, since I lacked the ability to smell anything for nearly four years until I finally underwent nasal surgery. How did I get my smell back? Because "physically" focused doctors realized that a series of specific non-cancerous polyps had developed in my nasal passages and their removal allowed for my sense of smell to return.

To ignore the tremendous progress science has made (because it focused on the physicality of vision, hearing, smelling, and touch) in the past century about how incoming stimuli engenders internal experiences in the brain is to create a false impression about the current state of neuroscience and its evolution.

Quite frankly, I feel our discussion is getting bogged down in metaphysical ultimacies when the core issue should be focused on the most viable pathways for understanding the nature of awareness and why it arises in homo sapiens the way it does.

This doesn't mean that in our pursuit of third person objective reports we have to somehow a priori exclude first person subjective experiences. To the contrary, they both go hand in hand, as any dentist and any Lasik surgeon knows.

Yet, Don Salmon seems stuck to a definitional dead zone that betrays what science has uncovered over the past few centuries. He writes, "So, when someone offers you a physicalist view, remember that he has just disappeared the entire universe—at least, the universe as we know it and live it—no color, all sights have vanished—inexplicable! No sounds, nothing solid or tangible ('solid' that is, in the sense of lived experience, not the physicist's definition of 'solid'). No smell, no taste. And since emotion as lived experience, and thoughts as lived experience, and comprehension as lived experience— well, as all cognition, affect and volition as lived experience—is admitted (at least, in dark corners of philosophically inclined academic journals) to be inexplicable—not only does the world disappear but you disappear as well."

This sweeping claim by Salmon, to put this in the politest of terms, is a classic straw man argument, since offering a physicalist view doesn't necessitate "disappearing the entire universe" of one's subjective experiences. Rather, by positing the physical correlations and constraints behind vision, smell, touch, and hearing, damaged senses can be repaired and new vistas reopened. I think Salmon is under the false impression that a physicalist understanding of experience somehow has to negate one's subjective experience of the same. Salmon also keeps repeating a mistaken canard when he says all of human subjective experience is "inexplicable" when, in point of fact and observation, nothing is farther from the truth.

As I have written previously, if seeing, hearing, touching, and smelling has a physical basis (which it obviously does), then is it really that much of a stretch for those interested in consciousness to focus on the physics of awareness by paying close attention to how our brains give rise to self-reflection? I think not, but Salmon is under the impression that such a quest is akin to searching for the Flying Spaghetti Monster and even goes so far as to write that the scientific work to ground consciousness in its material corpus is "quite a strange notion."

Strange? I think not, especially when we are all acutely aware of how tiny chemicals can dramatically alter our subjective experiences of the world at large. Indeed, to ignore the physics of awareness is to be blind to the very mechanisms that can both guide and disrupt our lives--from our inner ear equilibrium to nausea inducing vertigo.

Salmon underlines his purview when he writes that "physicalism is a faith-based, unfalsifiable, bizarre, non-empirical, dogmatic belief system which is impeding progress in virtually all areas of science, particularly neuroscience, but in evolutionary biology and physics as well."

But in making this dogmatic assertion Salmon once again indulges in a straw man argument, since science is essentially a practical process not a faith based one, and the idea of pursuing the physics of our combinatorial awareness isn't a journey to ontologically prove that all is mere matter. No, it is much simpler than that and much more transparent. It just happens to be the case that when studying the ins and outs of our anatomy, scientists have made tremendous progress when they grounded their hypotheses in the empirical arena. If scientists made greater advances by following ghosts and goblins, then they would follow that pathway instead. But they haven't and therein lay the key point that is lost here in our discussions.

For instance, I remember twenty years ago of having a truly awful tooth ache (my "first-person" experience of pain was nearly unbearable). Naturally, I tried to figure out the source of where "my experience of pain" arose. I wish I could say that my Ramana Maharshi influenced meditation sittings helped, but they did not. It was only after I went to a dentist (grounded as he was in the "physicalism" of pain via impacted wisdom teeth) who immediately realized that I had infected root canals and proceeded within all of but an hour to eradicate the source of my immense discomfort. I think I hugged him in the parking lot as we left, so thankful I was that he could be so masterful in alleviating me from my personal tooth torture.

Now most dentists don't spend their time in dental schools ruminating on the metaphysics of first person awareness vs. third person objectivity, because they take on the whole a much more practical and hands-on approach. They focus, in sum, on the material composition of teeth and their relation to the jaw and mouth. They also, and here we are thankful to live in the 21st century and not prior ones, pay close attention to minimizing the pain associated with dental work.

My argument is that science works precisely because it is practical. But that doesn't mean that in focusing on physical causes or correlations one has to somehow categorically exclude subjective experiences. Quite the opposite, since any doctor or dentist or neuroscientist worth his/her salt will be keenly interested in what one experiences under differing circumstances.

Hence, the notion that scientists are out to prove the ultimate material basis of all things, particularly consciousness, misses the essential praxis of what science is about. We are all scientists to some measure, as T. H. Huxley famously essayed a 150 years ago,

"The method of scientific investigation is nothing but the expression of the necessary mode of working of the human mind. It is simply the mode at which all phenomena are reasoned about, rendered precise and exact. There is no more difference, between the mental operations of a man of science and those of an ordinary person, than there is between the operations and methods of a baker or of a butcher weighing out his goods in common scales, and the operations of a chemist in performing a difficult and complex analysis by means of his balance and finely graduated weights. It is not that the action of the scales in the one case, and the balance in the other, differ in the principles of their construction or manner of working; but the beam of one is set on an infinitely finer axis than the other, and of course turns by the addition of a much smaller weight."

And in this pursuit, we are free to conjure up any hypothesis we wish about a given subject, but the linchpin in our theory making is that we are willing to air out our own ideas and have them competitively tested to see how well they hold up under scrutiny and rigorous experimentation. Thus, pursuing the physics of awareness doesn't automatically prevent any other budding scientist from following a different line of inquiry. It just happens to be the case (presently) that centering on the brain has produced amazing (and useful) results in better understanding a large range of different mental states.

This is why Huxley could argue that it is readily permissible to hypothesize that the moon is made of green cheese. The caveat, though, is that in order for such an idea to hold scientific credibility it must be rigorously vetted amongst other competing hypotheses. And in this winnowing pathway, the most viable overlay holds sway, if even only temporarily since science is always open to evolve and change when better ideas emerge.

As Huxley wisely wrote, "Do not allow yourselves to be misled by the common notion that an hypothesis is untrustworthy simply because it is an hypothesis. It is often urged, in respect to some scientific conclusion, that, after all, it is only an hypothesis. But what more have we to guide us in nine-tenths of the most important affairs of daily life than hypotheses, and often very ill based ones? So that in science, where the evidence of a hypothesis is subjected to the most rigid examination, we may rightly pursue the same course. You may have hypotheses and hypotheses. A man may say, if he likes, that the moon is made of green cheese: that is an hypothesis. But another man, who has devoted a great deal of time and attention to the subject, and availed himself of the most powerful telescopes and the results of the observations of others, declares that in his opinion it is probably composed of materials very similar to those of which our own earth is made up: and that is also only an hypothesis. But I need not tell you that there is an enormous difference in the value of the two hypotheses. That one which is based on sound scientific knowledge is sure to have a corresponding value; and that which is a mere hasty random guess is likely to have but little value. Every great step in our progress in discovering causes has been made in exactly the same way as that which I have detailed to you. A person observing the occurrence of certain facts and phenomena asks, naturally enough, what process, what kind of operation known to occur in Nature applied to the particular case, will unravel and explain the mystery? Hence you have the scientific hypothesis; and its value will be proportionate to the care and completeness with which its basis has been tested and verified. It is in these matters as in the commonest affairs of practical life: the guess of the fool will be folly, while the guess of the wise man will contain wisdom. In all cases, you see that the value of the result depends on the patience and faithfulness with which the investigator applies to his hypothesis every possible kind of verification."

In my science and religion classes at California State University, Long Beach and Mt. San Antonio College in Walnut, I often hear complaints from some of my more devout religious students (usually Christian and usually fundamentalist) that science takes the mystery away from the universe and tends to leave only a depressing and materialist residue, where the beauty and majesty of life is reduced to a Darwinian struggle over differential reproductive successes.

While at first glance this may seem to be a reasonable response to the tremendous progress science has made in explaining hitherto inexplicable phenomena, the truth is the opposite of what many fear. Science, due to its progressive and corrective nature, actually expands our aesthetic horizons and thereby exponentially increases our sense of wonder and mystery about the cosmos at large.

A good example of this is to juxtapose the latest findings in astronomy with the Hebrew description of creation as espoused in the first book of the *Tanakh* that was written down thousands of years ago. In Genesis Chapter 1, verses 14-19, it states, "God made two great lights—the greater light to govern the day and the lesser light to govern the night. He also made the stars. God set them in the vault of the sky to give light on the earth, to govern the day and the night, and to separate light from darkness."

Forgetting for the moment that the moon ("the lesser light to govern the night") receives its light primarily from the sun, the Genesis account, though poetic and beautiful, is hampered by a limited view of the cosmos. This is not surprising, of course, given that in those days there were no telescopes and no information about how planets and stars actually formed.

However, with the advent of astronomy, especially when it discarded some of its more superstitious elements (such as astrology), our understanding of the universe has undergone a radical expansion. As a quote from the movie, *Little Things that*

Jiggle put it, "Astronomy is Genesis rewritten and expanded nightly."

We are now keenly aware that there are over 100 thousand million stars in our Milky Way galaxy alone and almost daily we learn about newly discovered distant planets, black holes, and other stellar phenomena.

Astronomy's view of the universe, like the big bang that formed it some 13.8 billion years ago, is expanding at an unimaginable rate and along with it so is our understanding of the vastness that surrounds us.

Conversely, with the advent of high powered microscopes, we have the ability to see hidden worlds buried deep inside living cells, providing us with intricate details about atomic worlds within.

Thus, I invariably explain to my religious minded students that science doesn't take away the mystery and the sublimity of existence (even as it tries to explain much of it), but only enlarges our parochial perspectives by its ever-increasing illuminations on what has erstwhile been kept dark and hidden.

This is why Richard Feynman, the renowned physicist and Nobel prize winner for his work on quantum electrodynamics, got a bit miffed by his artist friend who criticized him and other scientists for diminishing the beauty and charm of nature by dissecting it.

As Feynman famously recounted in a filmed interview, "I have a friend who's an artist and he's some times taken a view which I don't agree with very well. He'll hold up a flower and say, 'look how beautiful it is,' and I'll agree, I think. And he says, 'you see, I as an artist can see how beautiful this is, but you as a scientist, oh, take this all apart and it becomes a dull thing.' And I think he's kind of nutty. First of all, the beauty that he sees is available to other people and to me, too, I believe, although I might not be quite as refined aesthetically as he is. But I can appreciate the beauty of a flower. At the same time, I see much more about the flower that he sees. I could imagine the cells in there, the complicated actions inside which also have a beauty. I mean, it's not just beauty at this dimension of one centimeter: there is also beauty at a smaller dimension, the inner structure...also the processes. The fact that the colors in the flower are evolved in order to attract insects to pollinate it is interesting -- it means that insects can see the color. It adds a question -- does this aesthetic sense also exist in the lower forms that are...why is it aesthetic, all kinds of interesting questions which a science knowledge only adds to the excitement and mystery and the awe of a flower. It only adds. I don't understand how it subtracts."

Richard Feynman astutely points out that science doesn't diminish our capacity for wonder or our appreciation for beauty, but only accentuates our aesthetic sense.

Those religions which believe only in *sola scriptura* are by their very definitions deductive systems of inquiry that continually look to reaffirm their a priori belief systems and disregard any inductive quest which could potentially upend or transform its core doctrines. This type of religion, in other words, kills the spirit of seeing knowledge by fearing the consequences of new information and thereby tragically (if unconsciously) snuffing out an appreciation for an unfolding mystery.

Feynman's flower is a multi-leveled metaphor that reminds us that the wide-eyed pursuit of science can only enrich our wonder of the multiverse in which we find ourselves.

As Feynman himself said best, "Poets say science takes away from the beauty of the stars - mere globs of gas atoms. I, too, can see the stars on a desert night, and feel them. But do I see less or more?"

7 | The Ivash Caution

I enjoyed Giorgio Piacenza's passionate admonitions to me [in "There's More Than Meet's the Eye"] about the multi-dimensional reality of E.T.'s. As he boldly proclaims (going where skeptics tend not to go): "There aren't only unidentified but genuine cases of interaction between non terrestrial intelligent beings that use a transdimensional technology combining natural physical reality and the Subtle Realm." But Piacenza doesn't merely stop there but goes on and explains, "Sometimes these interactive events maintain a 'high strangeness' of classical physics-defying behavior and require an enhanced understanding of how far the scope of 'nature' and 'physis' reaches. Sometimes there are clear, multiple witnessed, daytime, radar-tracked, objectively real observations with metallic-looking, structured craft. Sometimes, physical evidence of transformative contact experiences has been analyzed."

Given what he believes to be overwhelming evidence of E.T.I's (extra terrestrial intelligences), Piacenza very clearly states his thesis without any hesitation: "Let me tell you that there's more than meets the eye and that in many cases the extraterrestrial hypothesis is the most rational." Those last two words "most rational" underline precisely where Piacenza and I depart company.

While it is certainly true that the very basis of this planet is alien in origin and that our true conception lies at the heart of the big bang, it is not so certain that higher forms of intelligence have intervened (with apologies to Xenu, L. Ron Hubbard and to DC10 aficionados) in the day to day lives of humankind. I can appreciate that there will (and should be) rigorous debate over these kinds of issues and that given the vast, even if questionable, array of UFO reports from around the world we should be careful not to be too cynical or too closed-minded to occasionally entertain outrageous hypotheses on this subject.

In this regard, I think the skeptic and the believer can agree with Carl Sagan when he beautifully exclaimed, "We've begun, at last, to wonder about our origins. Star stuff, contemplating the stars organized collections of 10 billion-billion-billion atoms contemplating the evolution of matter tracing that long path by which it arrived at consciousness here on the planet Earth and perhaps, throughout the cosmos."

In this sense of wonder, I thought it might be useful to take a different route and go back nearly 7 decades and eavesdrop on one of the most celebrated, if not perfectly recalled, lunch conversations in the history of UFO theorizing. Enrico Fermi, the Nobel prize winning Italian physicist who did prodigious work on quantum theory and development on the first atomic bomb during the Manhattan Project, unexpectedly asked his colleagues (Emil Konopinski, Edward Teller, and Herbert York) over an informal lunch, "Where are they?" Although each participant has a slightly different recollection of what precipitated Fermi's query, they all agree that it was on the subject of space travel and the purported reality of flying saucers.

This in itself is not so surprising since there had been something of a media blitz (with the usual conspiratorial hysteria) about UFOs since Kenneth Arnold's 1947 flying saucer sighting. However, what was unusual was the inherent implication of Fermi's quip. Given the likelihood that there are suns and planetary systems much older than ours, replete with intelligent life forms, Fermi wondered why we had yet to see evidence of their existence. Surely if there were such advanced civilizations, presumably dating back millions of year before our own evolutionary advent, then "where are they?"

Later, in 1961 Frank Drake added to Fermi's query by proposing an equation about the probability of advanced alien interaction that is based on estimating the following:

N = The number of communicative civilizations

R* = The rate of formation of suitable stars (stars such as our Sun) fp = The fraction of those stars with planets. (Current evidence indicates that planetary systems may be common for stars like the Sun.)

ne = The number of Earth-like worlds per planetary system

fl = The fraction of those Earth-like planets where life actually develops

fi = The fraction of life sites where intelligence develops

fc = The fraction of communicative planets (those on which electromagnetic communications technology develops)

L = The "lifetime" of communicating civilizations

According to *setileague.org,* Drake currently guesstimates that there may be 10,000 or so "communicative civilizations in the Milky Way."

In contrast, though, in 1975 Michael H. Hart in a widely cited paper, "An Explanation for the Absence of Extraterrestrials on Earth" extends Fermi's query into a surprising and more radical conclusion, especially among UFOists: The reason we have no E.T. encounters is because highly intelligent alien civilizations don't exist. Because of this Hart believes that electromagnetic monitoring for E.T.'s is "probably a waste of time and money" and "in the long run, cultures descended directly from ours will probably occupy most of the habitable planets in our Galaxy."

Clearly, most of the established science community (though open to the possibility of intelligent life in the universe) is pretty much convinced that there is no overwhelming evidence to suggest that planet earth has already been touched by alien civilizations, even if they still remain quite open minded regardless of the unfavorable odds.

However, there are many others (including some scientists), as Giorgio Piacenza points out, that believe that planet earth has already had been visited. One of the most famous scientists in this pantheon was the late Francis Crick who proposed a variation of panspermia, where "small grains containing DNA, or the building blocks of life, could be loaded on a brace of rockets and fired randomly in all directions. Crick and Orgel estimated that a payload of one metric ton could contain 1017 micro-organisms organized in ten or a hundred separate samples. This would be the best, most cost effective strategy for seeding life on a compatible planet at some time in the future. The strategy of directed panspermia may have already been pursued by an advanced civilization facing catastrophic annihilation, or hoping to terraform planets for later colonization."--[*http://www.panspermia-theory.com/directed-panspermia/*].

A couple of decades later, however, Crick and Orgel changed their view and championed a more naturalistic explanation for the evolution of DNA. As they opined in *Anticipating an RNA World*, "We did not seriously consider the possibility that there was a midwife, a replicating pre-RNA world of quite different chemistry based, for example, on clays, as suggested by Cairns-Smith, or an alternative organic polymer. Such a pre-RNA world would have possessed the catalytic activity necessary to start the RNA world but it may not have needed to transfer its genetic information directly to that of the new (RNA) replication system. We now find this idea attractive."

With so many contrarian views, even from those skeptical of E.T.I., it is best to keep our options open to a wide variety of perspectives. Of course, we should also be doubly cautious since such unexplored terrain tends be an open season for all sorts of crackpot theories that only muddle an honest and reasonable discussion of the subject.

As Richard Feynman so wisely cautioned when speaking of things yet unproven, "Some years ago I had a conversation with a layman about flying saucers — because I am scientific I know all about flying saucers! I said "I don't think there are flying saucers'. So my antagonist said, "Is it impossible that there are flying saucers? Can you prove that it's impossible?" "No", I said, "I can't prove it's impossible. It's just very unlikely". At that he said, "You are very unscientific. If you can't prove it impossible then how can you say that it's unlikely?" But that is the way that is scientific. It is scientific only to say what is more likely and what less likely, and not to be proving all the time the possible and impossible. To define what I mean, I might have said to him, 'Listen, I mean that from my knowledge of the world that I see around me, I think that it is much more likely that the reports of flying saucers are the results of the known irrational characteristics of terrestrial intelligence than of the unknown rational efforts of extra-terrestrial intelligence.' It is just more likely. That is all."

Intriguingly, the subject of E.T.I. raises a very pertinent issue that underlines the sociology of knowledge itself. Isn't the debate between believers and skeptics predicated on the shifting sands of what one perceives/believes to be plausible? And isn't that very plausibility a socially mediated structure which can oscillate depending on cultural and generational contexts?

As Robin Phillips lucidly explains, "The phrase 'plausibility structures' was coined by sociologist Peter L. Berger to refer to the conditions in a society that make certain beliefs seem reasonable or unreasonable. Why is it that a proposition which, at one time and place, might seem completely self-evident and not even in need of argument, will seem totally absurd in another time and place? Questions like this force us to be attentive to more than merely what people believe, but the plausibility structures that explain why certain beliefs feel normal."

This explains, at least in part, why Giorgio Piacenza's views can be viewed both sympathetically (as in, "Okay, let's keep open to alternative channels of information") and dismissively, particularly when he makes such outrageous claims as "Let me tell you that some people involved occasionally begin to receive coordinates and other unique transmissions through automatic writing." UFO's is a troublesome enough field without adding channeling and automatic writing into the mix. Indeed, there are so many vying possibilities about E.T.I's that it is extremely difficult to differentiate what may be a potentially fruitful line of inquiry from a pseudo-scientific smattering of balderdash.

It is not surprising, therefore, that a serious researcher in this area may suffer from an acute case of informational vertigo. Perhaps as skeptics we shouldn't be too dismissive at this stage about borderline ideas or even nutty ones, particularly when as renowned a scientist as Francis Crick can seriously argue (during a period in his career) for specially designed exobiological rockets charged with viable DNA sent throughout space in order to seed future planets with life.

This reminds me of an old friend of mine Gene Ivash, now deceased, who I first met in North India in December of 1981. Gene was a theoretical physicist at the University of Texas and an expert on the mathematics underlying quantum mechanics. In the midst of our discussions, I was quite shocked to learn that this eminent physicist, well trained in the hardest of sciences, was an avid subscriber and reader of *Fate* Magazine, sometimes impolitely called the "National Enquirer of all things paranormal." Even though I had an article on the *Bhrigu Samhita* coming out in that very magazine within months, Gene noticed my obvious bemusement. So I asked him why he would read such a magazine given its somewhat questionable credentials. His answer was both wise and revealing. Gene politely responded that the best tool in any scientist's arsenal is his imagination and thus it is important not bury one's self too deeply in a rut by limiting one's reading material simply because it looks undignified. He then punctuated his purview with Einstein's famous quote,

"Logic will get you from A to Z [but] imagination will get you everywhere." Years later I would recall Gene's wise counsel whenever I got too comfortable within my own skepticism. I even gave it a meme and whenever I start wielding Occam's Razor somewhat drunkenly I remember "Ivash's Caution" and realize anew that timeless quote from Hamlet, *"And therefore as a stranger give it welcome. There are more things in heaven and earth, Horatio, than are dreamt of in your philosophy."*

I was brought up as a believer. My first indoctrination was within the Roman Catholic Church, where unbeknownst to me I was immersed in baptismal waters before I could even utter a coherent sentence. My parents, particularly my father, were regular weekly mass attenders, and me and my two brothers and one sister all attended Catholic schools from first grade until the end of high school. I received my first communion at the age of 7 and was genuinely moved by the ritual, so much so that our first grade nun, the wonderful Sister Susanna singled me out the next day and said that I was the only one out of some 40 or so students who truly appreciated the sacramental moment.

Although I was taught the Baltimore catechism, a near century old text of all things orthodox in Catholic dogma, I was never very good at mastering its arcane strictures. Indeed, in third grade we had a fairly vicious nun of the Blessed Virgin Mary order or BVM for short (which we mischievously translated as "black veiled monsters") who created an intellectual baseball game based on questions from the catechism. Answer an easy question (was Jesus the son of God?) and you advanced to first base. Answer a more difficult question, such as what is the Immaculate Conception? (Mary was freed from original sin even in the womb), and you got to round the bases for a home run. Needless to say, I never got to second base.

As a young devotee immersed in the life and work of Catholic saints (I chose my confirmation name from St. Francis Xavier, famous for his missionary work in India) I made a wager at the age of eight or nine with my childhood friend Pat Donahue that I would one day become a monk. Of course, that never panned out since I was far too attracted to the opposite sex to burden myself with a vow of celibacy.

However, at the age of ten and eleven I became quite interested in Eastern philosophy and religion and this, along with my mother doubting Pope Paul VI's *Encyclical Letter, Humane Vitae* (which condemned artificial birth control) jilted my rigid and parochial belief system. This was perhaps the beginning of my evolving skeptical attitude towards all things religious, particularly ideologies that demanded blind allegiance.

I was still deeply interested in the spiritual quest, though, and at the age of 15 I had a genuinely transformative experience when I spoke in tongues after a charismatic mass held at Loyola Marymount in Los Angeles, which I have written about in more detail elsewhere. The glossolalia moment opened me up to all things paranormal, even if I was resistant to ritualized calcifications.

I had a number of wondrous and (at the time) inexplicable experiences, ranging from a spontaneous, albeit partial awakening of my kundalini to a quasi near-death experience where I felt completely removed from my body and entered into (what I sensed was) another world.

In this regard, I was open to the interior quest and fully engaged myself in pursuing it. But at the same time I also developed a questioning attitude about the various interpretations that were almost invariably intertwined with such experiences.

For instance, even though I was mesmerized by speaking in tongues I rejected the Christian party line that it was a unique gift only bestowed on those who accepted Jesus as their personal Lord and Savior.

After doing the "breath of fire" exercise during an intense yoga session at the old redwood temple at the back of the Source restaurant in Hollywood I was quite bemused to actually feel a five pointed star of energy in my upturned left hand. Yet I resisted Father Yod's superluminal explanation and instead reflected that it was probably due to something psychosomatic.

This two-pronged approach of mine—being experientially engaged and skeptically analytical—occasionally gave me an epistemological headache. Nevertheless, it allowed me to witness some quite impressive, if not altogether paranormal, happenings. I was also a romantic of sorts and thus had a fondness for mystical literature, particularly tales that described enlightened states of awareness.

In the early 1980s I wrote a number of articles and book reviews for Fate Magazine, which was devoted to unexplained phenomena, such as UFOs, astrology, ghosts, clairvoyance, and all things bizarre. By the late 1980s I was even teaching a graduate course on parapsychology at the University of Humanistic Studies in Del Mar.

Yet, something snapped in me along the way, which made me wary of paranormal claims in general, and I changed from being a potential believer to a more hard lined skeptic. I am sure part of this transformation was due to my critical analyses of a number of new age religions and their leaders during my undergraduate and graduate studies. Metaphorically speaking, when I was younger I was like Dorothy in the *Wizard of Oz* and naively followed the yellow brick road and found the Emerald City and its designated leader enchanting, but when I got a bit older (and like Toto who pulled back the curtain on the Wizard only to find a balloonist from Kansas) and saw what went on behind the inflated scenery of the guru game, I became disillusioned with the hype that surrounded much of the transpersonal quest.

I also saw how easy it was to be duped by alleged miracle workers, whether it was Sathya Sai Baba's bad sleight of hand magic tricks or John-Roger Hinkins faking psychic knowledge when, in fact, he wiretapped his own house and eavesdropped on unsuspecting students.

I became suspicious of paranormal claims in general. I became jaded.

All of this, of course, may be chalked up to growing older and part of the natural progression of developing a more rationalist approach to the world at large. But being overly skeptical can devolve (as it too often does) into cranky cynicism that ends up cutting one off from new information and new vistas. Science is truly a wonderful pursuit but its shadow is scientism and sometimes one can conflate them, mistakenly believing they are one and the same when, in fact, they are polar opposites.

THE STAR TREK BRAIN METAPHOR

Hang out with naive believers long enough and you can start accepting all sorts of silliness as true, such as when a Christian preacher suggests that he would accept that 2 = 2 equals 5 if it was written in the Bible, since the Bible cannot be doubted. On the other hand, hang out with cynical skeptics long enough and you can too readily dismiss any new form of knowledge, since it doesn't fit into your model of how the universe should operate. A sad, but telling example of this kind of resistant mindset can be found in the old Soviet Union with Trofim Lysenko's adamant rejection of Mendelian genetics and how politics trumped open ended science.

The problem on either side of the knowledge equation is a simple one. How willing are we to be wrong? Or, put in a different way, what kind of evidence is sufficient to change our mind? Are we so set in our paradigmatic ways that we refuse to even consider alternatives that don't fit our favored models? The knife of reason cuts both ways and all too often we may find ourselves protecting our cherished ideals versus truly reconsidering what we believe in light of new and compelling findings.

Sam Harris in a recent debate with Daniel Dennett over free will captured the difficulty humans have in actually changing their minds when we are already deeply entrenched in a particular position. Cautions Harris, "In recent years, I have spent so much time debating scientists, philosophers, and other scholars that I've begun to doubt whether any smart person retains the ability to change his mind. This is one of the great scandals of intellectual life: The virtues of rational discourse are everywhere espoused, and yet witnessing someone relinquish a cherished opinion in real time is about as common as seeing a supernova explode overhead. The perpetual stalemate one encounters in public debates is annoying because it is so clearly the product of motivated reasoning, self-deception, and other failures of rationality—and yet we've grown to expect it on every topic, no matter how intelligent and well-intentioned the participants."

Jonathan Haidt, ironically, wrote an essay entitled "Why Sam Harris is Unlikely to Change his Mind" based on a questionable analysis of how many times "certainty" buzz words show up in Harris' two books, *The End of Faith* and *The Moral Landscape* [For Harris' response, see "The Pleasure of Changing My Mind"]. Haidt found that "Of the 75,000 words in *The End of Faith*, 2.24% of them connote or are associated with certainty. (I also analyzed *The Moral Landscape*—it came out at 2.34%.)"

Haidt claimed that the percentage of "certainty" words (such as "always", "never", "undeniable", etc.) in Harris' two tomes was significantly greater than what he found in Ann Coulter's *Treason*, Sean Hannity's *Deliver Us From Evil* and Glenn Beck's *Common Sense*.

Based upon this shaky scaffolding, Haidt confidently wagered a 10,000-dollar bet that Sam Harris would not change his mind about his thesis in *The Moral Landscape*. As Haidt explains, "In the opening paragraph of his *Enquiry Concerning the Principles of Morals*, David Hume described the futility of arguing with people who are overly certain about their principles. He noted that 'as reasoning is not the source, whence [such a] disputant derives his tenets; it is in vain to expect, that any logic, which speaks not to the affections, will ever engage him to embrace sounder principles.' If Hume is right, then what is the likely outcome of *The Moral Landscape Challenge*? What are the odds that anyone will change Harris's mind with a reasoned essay of under 1000 words? I'll put my money on Hume and issue my own challenge, The Righteous Mind challenge: If anyone can convince Harris to renounce his views, I'll pay Harris the $10,000 that it would cost him to do so."

What Haidt is pointing out (even if we disagree with his methodology) is that human reason is a flawed instrument and while we may think we are being tolerant and reasonable other factors (including deep seated emotions and biases) come into play and may sway us blindly in directions we don't expect.

I know in my own life that what I found convincing and evidential at one age (at 17, say) seemed much less so at another age (50, say). The real culprit here may well be our confused evolutionary history, where our brains are more or less jerry rigged with not always complementary agendas. While the triune brain analogy may be overused and not altogether accurate, it does provide us with a glimpse of why we behave in such contrarian ways. Employing a metaphor from *Star Trek* and cribbing Edward O. Wilson's terminology from *Consilience*, our brain is one confused enterprise indeed: The rational part of ourselves--the cerebral cortex--is Spock or Data like, where logic and reason holds court (what Wilson calls "heartless"); the emotional part of ourselves--the midbrain is Captain Kirk (what Wilson calls "heartstrings"); and the brute physical part of ourselves--the reptilian brain stem--is Scotty, the engine master (what Wilson calls "heartbeats"). Granted that the *Enterprise* metaphor is a bit silly and not altogether accurate, but there is one telling point that holds up: Captain Kirk drives the ship, just as our emotional selves are the driving forces in our own lives. We may wish to be purely logical or Spockian in our views, but the truth is that we are a combinatorial creature with competing forces at play.

I think Errol Flynn, the famous movie actor of the 1930s/1940s and the raucously funny author of *My Wicked Wicked Ways*, summed up the human dilemma quite nicely when he opined about himself that he was "an octagon of contradictions which may in itself be no contradiction."

I fondly remember my old Ph.D. advisor, Professor Bennett Berger from UCSD, who used to chuckle after reading how authors responded to scathing reviews of their texts in the *New York Review of Books*. Berger noticed that rarely, if ever, did an impugned author ever admit that the reviewer was perfectly justified in his critique. Rather, such authors write long-winded rejoinders explaining how they were wrongly mistreated.

It isn't easy to switch allegiances, especially if one has invested lots of time and energy in a particular purview. Instead, we tend to indulge in all sorts of ideological work to reconcile whatever gaps may exist between our pet model and contrarian findings.

Perhaps the jetty between those who profess to be skeptics and those who profess to be believers isn't as wide and jagged as we might at first suspect. Perhaps the impasse has more to do with how interested and passionate we are in contravening our cherished inclinations.

Once we become settled and comfortable in our positions, how often do we venture out and spend serious time trying to upend what we hold to be true? If we are creationists, how much time and effort do we give to truly grasping molecular and evolutionary biology? Conversely, if we are evolutionists, how much attention and serious study to we give to works espousing intelligent design?

As Jonathan Haidt pointedly explains, "In the 1980s and 1990s, social psychologists began documenting the awesome power of "motivated reasoning" and the "confirmation bias." People deploy their reasoning powers to find support for what they want to believe. Nobody has yet found a way to 'debias' people—to train people to look for evidence on the other side—once emotions or self-interest are activated. Also in the 1990s, the neuroscientist Antonio Damasio showed that reasoning depends on emotional reactions. When emotional areas of the brain are damaged, people don't become more rational; instead, they lose the ability to evaluate propositions intuitively and their reasoning gets bogged down in minutiae."

I have also noticed that in my own online debates with various posters over the years that there was some issues (usually related to theological claims) which really got under my skin and which brought out more piss and vinegar in my ripostes', whereas other topics left me somewhat cold and neutral and which allowed me to think more freely and loosely about the argument. In other words, intellectual debates can often sink into turf wars that have more to do with protecting marked territory and less to do with discovering what is true or evidential.

All of this came into sharp focus for me during the past month when I was asked by Alex Tsakiris to be interviewed for his *Skeptiko* podcast/website concerning my essay, "The Material Basis of Near-Death Experiences: Exploring the Patricia Churchland and the Alex Tsakiris Controversy" which was published on Frank Visser's *Integral World*.

To prepare for the conversation I spent time going through a number of Alex Tsakiris' previous interviews, particularly those that disagreed with his pro parapsychology perspective. I also spent a fairly large amount of time reviewing current books and articles on Near-Death experiences. Back in the late 1970s and early 1980s I read everything I could on this subject, since Raymond Moody's bestseller, *Life After Life*, opened up the field to a wider audience. I even had the occasion of hearing personally from NDE survivors and I was intrigued (and sometimes bemused) by what they recollected.

One of my high school students at Moreau back in 1979 talked privately to me after class and said that he too had an NDE but was reluctant to tell anyone about it because he didn't see Jesus or a religious figure. Instead, he told me that he went through a long dark tunnel and at the end in a brilliant scintillating light he saw the most beautiful bicycle ever. The moment he saw the bike he remembered that his birthday was in a couple of days and that his Mom had promised to buy him a Schwinn ten speed. That very thought brought him immediately back to his body and he regained consciousness in the emergency room. Later he was told that he had died on the operating table but was resuscitated minutes later.

I must confess that I was surprised to learn how much the field of NDE's has grown over the past few decades and how affirmative a number of doctors and scientists have been about the trans-neuronal nature of such experiences. The skeptic in me has tended over the years to see NDEs as part and parcel of how consciousness as a virtual simulator will project anything from our unconscious to encourage us to stay alive. This made sense to me since the one common denominator in all NDE recollections is that the patient did not die but lived long enough to report such transcendental happenings.

Yet, the more I read and studied NDE literature the more I realized that my own pet theory was in the minority, and that a sizable number of NDE researchers were convinced that near-death experiences were not merely the result of cerebral anoxia or some other brain machination. Indeed, many of these researchers (some with quite impressive credentials) were fully convinced that consciousness survives death and that a purely materialist explanation was insufficient to explain such numinous epiphanies.

While I still have a persistent bias that those favoring mysticism and parapsychology are better served by skeptics than they might at first suspect, I realized that there were serious researchers who were punching holes into my self as brain compound and that it might be wiser to simply listen to what they had to say instead of prematurely theorizing their findings away.

This, of course, doesn't mean that I have let go of neurophilosophy and the like, but only that science works better when we take seriously those ideas which upend what we ourselves believe is true. To accomplish this lofty goal, we have to always be open to being wrong, even if goes against our genetic grain.

I don't want to be wrong, but in the past 57 years on this planet I know all too well how often I have been mistaken about so many things. Yes, no doubt, there is much in parapsychology and the like that is nonsense, but the same can be said as well about some aspects of science, especially in fields like medicine which are still in their infancy.

The good news is that this is an exciting time since the Internet has allowed more voices to be heard than ever in the history of humankind and thus we live at a moment where information is not the sole province of only priests or an elite oligarchy. While I think it is altogether wise to be doubtful of extraordinary claims, I think it is also prudent to be skeptical of our own skepticism.

As Richard Haidt confessed at the end of his critique of Sam Harris, "If we want to improve our politics and our society, let's be reasonable about reason and its limitations. Of course, I have used my powers of reasoning (and intuition) to write this essay, and I have drawn on scientific studies to back up my claim that Harris is unlikely to change his mind and renounce his claims about morality. But people are complicated and it's always hazardous to use scientific studies to predict the behavior of an individual. I could well be wrong."

About the Authors

Andrea Diem-Lane is a Professor of Philosophy at Mt. San Antonio College. She received her Ph.D. and M.A. in Religious Studies from the University of California, Santa Barbara, where she did her doctoral studies under Professor Ninian Smart. Professor Diem received a B.A. in Psychology with an emphasis on Brain Research from the University of California, San Diego, where she did pioneering visual cortex research under the tutelage of Dr. V.S. Ramachandran. Dr. Diem is the author of several books including an interactive textbook on religion entitled *How Scholars Study the Sacred* and an interactive book on the famous Einstein-Bohr debate over the implications of quantum theory entitled *Spooky Physics*. Her most recent book is *Darwin's DNA: An Introduction to Evolutionary Philosophy*.

David Christopher Lane is a Professor of Philosophy at Mt. San Antonio College and an Adjunct Lecturer in Science and Religion at California State University, Long Beach. He received his Ph.D. in the Sociology of Knowledge from the University of California, San Diego, where he was also a recipient of a Regents Fellowship. He has taught previously at Warren College at UCSD, the University of London, and the California School of Professional Psychology. He has given invited lectures at various universities, including the London School of Economics. He is the author of a number of published books such as The *Making of a Spiritual Movement: The Untold Story of Paul Twitchell and Eckankar; The Radhasoami Tradition: A Critical History of Guru Succession; Exposing Cults: When the Skeptical Mind Confronts the Mystical;* and *The Unknowing Sage: The Life and Work of Baba Faqir Chand,* among others.